Narrow-Band Phenomena—Influence of Electrons with Both Band and Localized Character

W0115093

NATO ASI Series

Advanced Science Institutes Series

A series presenting the results of activities sponsored by the NATO Science Committee, which aims at the dissemination of advanced scientific and technological knowledge, with a view to strengthening links between scientific communities.

The series is published by an international board of publishers in conjunction with the NATO Scientific Affairs Division

A	Life Sciences	Plenum Publishing Corporation
B	Physics	New York and London
C	Mathematical and Physical Sciences	Kluwer Academic Publishers
		Dordrecht, Boston, and London
D	Behavioral and Social Sciences	
E	Applied Sciences	
F	Computer and Systems Sciences	Springer-Verlag
G	Ecological Sciences	Berlin, Heidelberg, New York, London,
H	Cell Biology	Paris, and Tokyo

Recent Volumes in this Series

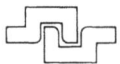

Series B: Physics

Narrow-Band Phenomena— Influence of Electrons with Both Band and Localized Character

Edited by

J. C. Fuggle

University of Nijmegen
Nijmegen, The Netherlands

G. A. Sawatzky

University of Groningen
Groningen, The Netherlands

and

J. W. Allen

The University of Michigan
Ann Arbor, Michigan

Plenum Press
New York and London
Published in cooperation with NATO Scientific Affairs Division

Proceedings of a NATO Advanced Research Workshop on
Narrow-Band Phenomena—Influence of Electrons with
Both Band and Localized Character,
held May 31–June 5, 1987,
in Staverden, The Netherlands

Library of Congress Cataloging in Publication Data

NATO Advanced Research Workshop on Narrow-Band Phenomena—Influence of
Electrons with Both Band and Localized Character (1987:Staverden, The
Netherlands) Narrow-band phenomena—influence of electrons with both band
and localized character.

(NATO ASI series. Series B, Physics; vol. 184)
"Proceedings of a NATO Advanced Research Workshop on Narrow-Band Pheno-
mena—Influence of Electrons with Both Band and Localized Character, held May
31—June 5, 1987, in Staverden, The Netherlands"—T.p. verso.
"Published in cooperation with NATO Scientific Affairs Division."
Includes bibliographical references and index.
1. Energy-band theory of solids—Congresses. 2. Free electron theory of metals
—Congresses. I. Fuggle, J. C. II. Sawatzky, G. A. III. Allen, J. W. (James Ward) IV.
North Atlantic Treaty Organization. Scientific Affairs Division. V. Title. VI. Series:
NATO ASI series. Series B, Physics; v. 184.
QC176.8.E4N375   1987   530.4′1   88-25250

ISBN-13: 978-1-4684-5561-8   e-ISBN-13: 978-1-4684-5559-5
DOI: 10.1007/978-1-4684-5559-5

© 1988 Plenum Press, New York
Softcover reprint of the hardcover 1st edition 1988
A Division of Plenum Publishing Corporation
233 Spring Street, New York, N.Y. 10013

All rights reserved

No part of this book may be reproduced, stored in a retrieval system,
or transmitted in any form or by any means, electronic, mechanical, photocopying,
microfilming, recording, or otherwise, without written permission from the Publisher

PREFACE

The understanding of electronic behaviour in solids when (some of) the valence electrons have both localized and band-like characteristics is one of the central problems of physics and chemistry in the second half of this century. Many advances have indeed been made using highly sophisticated techniques and concepts. Our objectives in bringing together specialists from different areas was cross-fertilization of ideas and redefinition of bottlenecks and problems. The testimony of the participants and the book which follows indicate a fair degree of success.

This book is a record of discussions aimed at digestion and reassessment of some of the recent major advances in our understanding of narrow bands. Note that we expressly asked participants to give a short readable account of the major problems in their field and not to emphasize their latest results to be as "technical" as they might be in a normal scientific article. We did not ask for complete reviews of what was going on in the field and this book should not be read as such. Neither should it be approached as the sort of educational text which the NATO ASI proceedings are supposed to be. We have tried to produce a useable account of a workshop in which an attempt was made to define real problems and to distinguish them from illusory problems.

The main requirements we tried to meet in organizing this meeting were:
1) 30-40 high level scientists who are not only talkative, but also good listeners. It is neither necessary nor desireable to have everybody in the field.
2) A good and relatively isolated location.
3) A reasonably well-defined concept and list of problems.
4) A fair proportion of people (the more the better) with whom the concept of the workshops has been extensively discussed and who are agreed on its aims.
5) A pretty robust sense of humor.

If there was a disappointment in this workshop it was that we didn't get further, and that it is still so difficult to fully appreciate the concepts of our colleagues. Nevertheless we were able to agree that some problems were essentially solved. This is important because a lot of time can go down the drain if half the world is trying to convince the other half of what the latter already believes. It would be good to organize more workshops of this sort from time to time to decide what are real problems. In particular we believe in documenting the results of such workshops.

As a last point in this preface we would like to thank all those who contributed to the success of the Narrow Band Workshop. These include the organization which supported the effort: the NATO advanced research workshop program. The Dutch Academy of Science, and the universities of Nijmegen and Groningen. Also we especially thank the speakers, chairmen, discussion group leaders, and all participants and all members of our research groups who contributed to the meeting. Finally, we note the burden of all the administration, correspondence and typing that fell to our secretaries, Desiree van der Wey and Anita Severijnse and stress that without their help nothing would have been possible.

J.C. Fuggle
G.A. Sawatzky
J.W. Allen

CONTENTS

C. HIGH ENERGY SPECTROSCOPIES OF NARROW BAND MATERIALS

D. HIGH TEMPERATURE SUPERCONDUCTORS

INTRODUCTION TO THE BOOK

1. MOTIVATION

The term "narrow band phenomena" is not absolutely defined but we can explain
the field covered by this book as follows: the electrons in an atom or
solid are in states characterized by a wave function and an energy level.
These states are sub-divided into the valence levels, whose wave functions
overlap in a compound, and the core levels, which are confined to the core
regions of the atom. In narrow band materials there are partially filled
states with characteristics between those of the localized core states and
the (normally) delocalized and weakly correlated, valence states. For these
systems a good understanding of the interesting physical properties requires
a good, detailed determination of the electronic structure, which is not
solely dominated by either band structure effects or by atomic correlations,
but is strongly influenced by both.

We loosely define narrow band phenomena as those arising when the
partially filled states of a condensed matter system have characteristics
intermediate between those of the limiting cases where either the band
structure, or the atomic correlations dominate. The theoretical treatment
of such a situation is one of the central scientific problems of the last
third of this century[1-8]. Even the problem of a single atom with strong
atomic correlations, embedded and interacting with a simple one-electron-
like metal is a problem which has not been exactly solved for even the
simplest possible hamiltonians. An exact solution looks even more remote
for the much more complicated problem of d- or f-impurities, for which the
atomic correlations give rise to a large number of multiplets, different for
each electronic configuration. Metals and compounds, in which some of the
component atoms have characteristics which are strongly influenced by the
strong electron-electron correlation, pose an even more difficult problem
because one now has a lattice of interacting, strongly correlated, atoms.

The scientific importance of understanding narrow band materials can
be appreciated when one remembers how widespread is the occurrence and the
range of the phenomena involved. Narrow bands may occur in both insulators
and metals and their consequences may be different in these two classes.
We would argue that almost all ferro-, antiferro-, and ferrimagnetic
materials involve narrow band effects. The most obvious narrow bands are
the 4f and 3d bands in lanthanides and transition metal compounds
respectively. Here one can often describe the material properties very

well by starting from an atomic picture and treating hybridization or mixing of states as a small perturbation. It is this perturbation that leads to transferred hyperfine interaction, the Anderson theory of superexchange, and the very successful Goodenough-Kanamori rules relating superexchange interactions to d-orbital filling and bond angles. In the rare earths one may easily rationalize these phenomena in terms of the wave functions of the partially filled, 4f, shell which are small compared to the sizes of the atoms so that overlap with the states on neighbouring atoms is small and one may say that the electron hopping integrals between the atoms are small. Indeed, in some of the heavy rare earths the hopping integrals are so small that one would like to neglect them altogether and say that the band aspects may be neglected completely. However, even here there is a problem because one needs some form of hopping to explain the magnetic interactions between the atoms, be it via superexchange in insulators, or a RKKY-type interaction in metals[9].

The transition metal d wave functions are small, but not as small compared to the atomic dimensions as are the 4f wave functions. Nevertheless nature has another way of producing very narrow bands in transition metal compounds in which the nearest neighbours of the cations are only anions and the cation-cation distances are very large. The only mechanism for "banding" of the cation levels is then via an excited state of the anion. If the covalent mixing of the cation and anion is given by t/Δ, where t is a transfer integral and Δ is a charge transfer energy, then the effective cation-cation hopping integral, (b) is of the order of t^2/Δ. For transition metal halides Δ is typically 5 eV and t is of the order of 1 eV, yielding band widths considerably less than 1 eV [10].

The phenomena associated with these narrow bands are very varied. One may list such examples as Mott-insulators and metal-insulator transitions of the Mott type[1-3]. Magnetic polarons[11], magnetic semiconductors[12] and the so-called mixed valence or inter-configuration fluctuation systems in which non-integral numbers of valence electrons lead to anomalous lattice constants and other thermodynamic properties[13-15]. In metallic systems with narrow bands there are a large number of anomalous low temperature phenomena. One example is the Kondo effect, which gives rise to a minimum in the electrical resistivity at about the "Kondo temperature" for simple metals with small amounts of magnetic impurities[8,16,17]. Another group involves the large linear term in the specific heat, and related magnetic behaviour, observed in heavy fermion materials at low temperatures[17-19]. Yet another is the problem of competition between hybridization and correlation, which leads to first ferromagnetism in the Stoner model, and finally local moment formation[7] as the correlation is increased with respect to the hybridization, or band width. Finally we may add that the ideas being developed for strongly correlated systems in ordered solids and impurity systems are also relevant for the localization of electrons in amorphous glasses[20].

It may be guessed from the above selection of the huge variety of narrow band phenomena and the wide range of their occurrence that they should be of considerable technological importance. No statistics are available but we believe this to be the case. Of course the variety of behaviour is far from fully explored and we still have insufficient insight to predict where interesting phenomena will occur (heavy fermions and the new discovery of superconductivity in copper perovskites are adequate illustrations of this point). But we can still note the new applications of phenomena we associate with narrow bands, in such diverse fields as catalysis, doped semiconductors and interfaces, magneto-optics, and even the "old" phenomena of magnetism (it is sometimes claimed that the value of magnets in modern computers exceeds the value of the semiconductors). These applications are of enormous economic importance in relation to the cost of our research.

SOME REMARKS ON THE ELECTRONIC STRUCTURE OF SOLIDS

To put all the effects described above into context we give a very short summary of some models of solid state electronic structure. For more details we refer the reader to the excellent books by Cox[21] and Mott[2]. As mentioned above the degree of electron localization ranges from the de-localized free electron gas to the atomic-like 4f orbital behaviour of most lanthanides. The most popular description of the electronic structure of solids for the physicist[22,23] is the free electron or nearly free electron like model in which the variation in potential due to the atoms in the solid is assumed to be negligible. Here the density of states varies as the square root of the energy, E. The electronic specific heat is $1/3\pi^2 N(E)k^2 T = T$, which is small, gives a linear rise with temperature, and is of the order of 1 mJ/K. The free-electron-like band theory predicts temperature-independent (Pauli) paramagnetic behaviour. The theory is assumed to work best for the alkali metals, although even for these metals spin and charge density wave ground states have been proposed[46] and there are indications that theoretical band widths are considerably too large[24].

A good criterion for the success of the free electron model is that the radial extent of the atomic wave function be large compared to the interatomic dimensions. A good compilation to judge the behaviour of solids in this respect is figure 1, prepared by van der Marel[28] and to describe it we plagiarize his text (with permission). "The figure is a plot of the ratio of the d and f volume to the Wigner Seitz volume $(r_1/r_{ws})^3$, where the r_1 (l=d or f) values were based on tabulated $\sqrt{<r^2>}$ Hartree-Fock or Dirac-Slater values for the elements[25-27] with minor modifications. The Wigner-Seitz radii were obtained from the elements in their common crystal phases[22,26,29]. We see in Figure 1 that all the values are smaller than one, indicating that on the average the d and f clouds lie inside the Wigner-

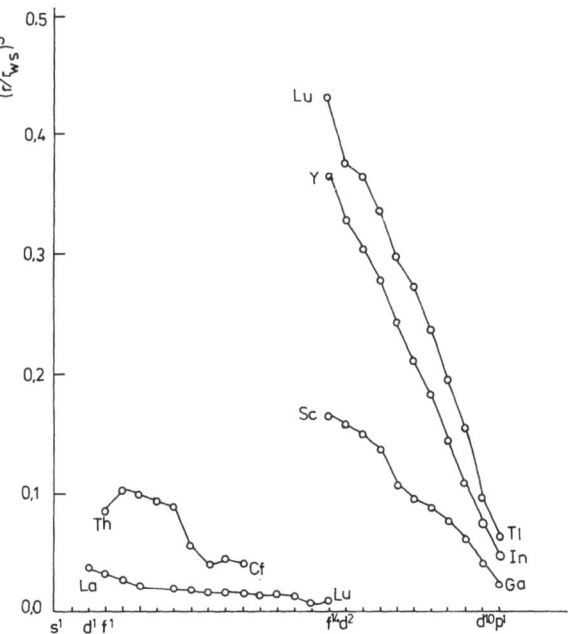

Figure 1. Ratio of 1 shell "volume" to Wigner Seitz volume of the 3d, 4d and 5d elements (l=2) and 4f and 5f elements (l=3) as a function of their position in the periodic system.

Seitz cell. Although this is no direct measure of overlap integrals, it is clear that elements with similar (r_1/r_{WS}) values have comparable band widths. Quite clearly, Am, Cm, Bk, and Cf are more confined than the 3d elements up to nickel, and more or less comparable to the lanthanides. From the spatial point of view, the lighter actinides belong to the same class as Mn, Fe, Co and Ni. The lighter 3d-transition metals are in the same way comparable to Au, Pd, Rh and Pt. A good estimate of the band widths as a function of interatomic distance can be obtained from Harrison's[47] scheme relating hopping matrix elements (t) between d states to the d state radii and the interatomic distance $(t\alpha(r_d)^3/R^5)$. So far we have been discussing the pure elements, but Figure 1 also gives a rough guide to the size effect in the overlap of d and f states with states centred on neighboring sites, and hence of hybridization between the d or f levels and the conduction bands.[48]

For solid state physicists the extreme opposite pole to free-electron-like behaviour is that of good (simple) insulators with an even number of electrons per unit cell. Here one can speak of filled valence bands with a gap before the onset of the conduction bands. In many ionic insulators, such as KCl one may treat the valence band states being situated on one atom (Cl) and the conduction states as being situated on the other (Na). The activation energy to conduction is then predominantly of "charge transfer" character. For a covalent insulator, like diamond or silicon, one would postulate a bonding-antibonding interaction to explain the insulation and one should speak of a "hybridization gap". In both limiting cases it is important for there to be an even number of electrons per unit cell for the material to be an insulator.

The description based on Wilson's [30] (1931) band theory predict metallic behaviour for systems with an uneven number of electrons per unit cell. One of the earliest problems to appear, after the first triumphs of band theory in the 1930's, was that of materials like CoO which, as pointed out by de Boer and Verwey[49], are insulators, despite the fact that they have an odd number of electrons per unit cell. Mott[1] first proposed a solution to this problem, based on the concept of large local electron-electron interactions which could cause a correlation gap to open. This concept is often explained to physics students using the idea of an imaginary lattice of hydrogen atoms separated by the large distance. Students are familiar with the idea that the ionization energy (E_I) of a hydrogen atom is 13.6 eV and that the electron affinity (E_A) is only 0.7 eV. They can thus accept the idea that the imaginary lattice is a Mott, or Mott-Hubbard insulator with a barrier to conductivity of 13.6 - 0.7 = 12.9 eV. In this case 12.9 eV would also represent the "effective Coulomb correlation energy", $U = (E_I-E_A)$, of the electrons. It is quite understandable, as in figure 1, that the width of the hydrogen 1s bands increases as the distance between the hydrogen atoms is decreased. This, together with a decrease of U because the polarizability varies as $1/R^4$ and decreased R produces increased screening, decreases the energy barrier to metallic behaviour.

An important development in the field of narrow band phenomena was the suggestion by Mott that there should be a sharp phase transition from insulating to metallic behaviour when some substances are heated, as for example in VO_2 and V_2O_3. In fact, as stated by Wilson[31], true Mott transition are rather rare, but Mott insulators are very important, constituting about half of all binary compounds of 3d transition metals[31] and being also important for thinking about the properties of many more 4f and 5f compounds.

In recent years we have come to realize that band gaps in transition metal compounds are probably seldom to be described in terms of any one of "charge transfer energies", "hybridization gaps" or "effective Coulomb correlation energies" but that all three effects play a role[32,21]. There

4

is increasing effort to identify the relative importance of these parameters, either by relating them to the parameters that can be measured spectroscopically, or by first principles computations. It is important to point out that many theoretical treatments of solids neglect the orbital degeneracies and especially the multiplet splittings, primarily for computational convenience. However multiplet splittings can be of the order of 10 eV in the rare earths, and 5 eV in transition metals, both of which are large compared to band widths in narrow band materials. This neglect can cause serious discrepancies when comparing various spectroscopies and ground state properties. Also the hybridization itself can be strongly multiplet dependent, both because of the large multiplet splitting as well as the symmetry dependence of transfer integrals.

It is logical at this point to mention a consequence to the electron correlation energy in systems where a Mott gap is not opened. The most important of these is the suppression of charge fluctuations in a solid. In a one electron description of the electronic structure of, for instance 3d transition metals, the occupation number of a 3d atomic orbital is assumed to be determined purely statistically, with the result that there is a finite probability of an atom to have from zero to ten electrons. Electron correlation strongly suppresses these variations and enhances the contributions of the configurations with d counts closest to the average number of 3d electrons. In extreme cases the fluctuations may be suppressed to such an extent that the electrons may be considered localized. In the 1950's and 60's there were many arguments about whether the magnetism of transition metals should be regarded as "itinerant" (band like) or localized. Nowadays it is generally agreed that the magnetism of transition metals is basically intermediate in character, that one can certainly explain lots of properties of transition metals using band theory, and that it would not be wise to rekindle old arguments. These issues are not central to this book but are background knowledge. However a related point which continually crops up in discussions on narrow band, and especially rare earth, systems concerns the occupation of the "shallow core" levels in the solid, such as the 4f levels. Terms like "mixed valence", "valence fluctuation" and "covalency" are often used in an effort to point out that the occupation of such an orbital need not be integral.

"Valency" is itself a concept with a confused history.[33] In chemical usage the valency of an element was defined in English around the beginning of this century. Initially it was the number of hydrogen atoms with which an element could combine, or for which it could substitute. Later it became twice the number of oxygen atoms with which an atom could combine (for a metal) or twice the number of atoms for which a metal could substitute. The concept of variable valency is quite old and was necessary because, for instance, transition metals could form compound of variable stoichiometry. Some lanthanides have compounds whose stoichiometry and crystal structure can only be explained in terms of atoms with more than one valency. E.g. Eu_3O_4 must contain Eu^{2+} and Eu^{3+}, and as these are on crystallographically distinct sites one may speak of "inhomogeneous mixed valence". On the other hand Eu_3S_4 also contains Eu^{2+} and Eu^{3+} but in the high temperature phase these are on crystallographically equivalent sites.

Of more fundamental importance is the phenomenon of homogeneous mixed valence which has been put forward for some compounds of Ce, Eu, Sm, Tm and Yb[34], although as will become clear later it may be desirable to consider at least Ce separately. "Mixed valence" affects many physical properties but is easily explained in terms of lattice constants. If we consider n to represent the position of a rare earth element (La=0, Ce=1, Pr=2, --- Yb=13, Lu=14) then the $4f^n$ configuration is associated with a three-valent ion. There are then three "valence" electrons. A $4f^{n+1}$ or $4f^{n-1}$ configuration will then be associated with 2 or 4 valent compounds, respectively[35]. In metallic compounds an increase in the number of valence electrons results in increased cohesion with consequent decrease in the lattice constant. The latter may thus be used as a monitor of the number of 4f electrons.

In general the lattice constants of the rare earth metals follow a simple curve as a function of atomic number, Z, with a gradual contraction as Z increases (the "lanthanide contraction", see figure 1). There are jumps in the lattice constant curves where a $4f^{n+1}$ or a $4f^{n-1}$ configuration occurs. However, some compounds exhibit a lattice constant which can only be explained if there is a non-integral f count on every rare earth site, or if the f electrons are contributing significantly to the cohesion of the solid. In the former case one speaks of "mixed valence" (dropping the explicit reference to the homogeneity) or "interconfiguration fluctuation". Until not so long ago, mixed valence was the only explanation considered to explain the anomalous lattice constants of rare earth materials, but since the early 1970's there has been increasing realisation of the possibility that the 4f levels could play a direct role, especially for Ce[36-39].

The idea of mixed valence, in its simplest form, implies that there must be a partially filled f band or level at, or very close to, the Fermi level in order to allow transfer of electrons between the 4f levels and the delocalized bands. Of course such a very narrow band at or near the Fermi level, has very profound consequences for the ground state properties of a solid. Once this was recognized, it led to a phase in which mixed valence was extremely fashionable and nearly every observation of an anomalous property of a rare earth material was likely to be followed by a claim of a new observation of mixed valence. It is easy to understand that a sharp level or band at the Fermi level can alter properties of a solid, such as specific heat, electrical resistivity, and magnetic moments. However, it must be recognized that anomalously high low-temperature specific heats and electrical resistivities may be indicative of more than just a narrow band of one-electron states at the Fermi level. Consider, for instance, the electrical resistivity of metals. Because the scattering due to phonons drops with T^5 it is negligible at low temperature. The residual resistivity at low temperatures is thus controlled by impurity scattering which is independent of temperature for non magnetic impurities. Now it has been known since 1930 that alloys with a magnetic impurity show a minimum in their resistivity (see fig. 2) which depends only weakly on the concentration of impurities[8,22,40,41] and that a rise in resistivity is found at low temperatures. It was only in 1963 that Kondo showed that the exchange interaction between the conduction electrons and the local moment of the impurity leads to scattering events in which the electronic spin of the conduction electron is flipped, with a compensating change in spin of the local moment. The Kondo effect is dependent on a sharp cut-off of the conduction wave vector distribution so that the divergence of the magnetic scattering distribution disappears at higher temperatures where the thermal distribution at the Fermi edge is rounded. The Kondo effect is a subtle and complex effect but there is a widespread feeling[39] that, through the work of people like Gunnarsson, Schönhammer, Allen and Martin[38,42], we are slowly beginning to get more insight into the Kondo problem.

It is very important to note that the observation of these low temperatures and low energy effects does not imply that the magnetic impurity has an electronic state at the Fermi level, i.e. the first ionization energy of the magnetic impurity is not necessarily (and seldom is) equal to the ionization energy for an electron at the Fermi level. This is a situation frequently met in other areas of "narrow band phenomena" where the term "Kondo" may be less appropriate.

Moving on in time, one of the more spectacular new phenomenon to be

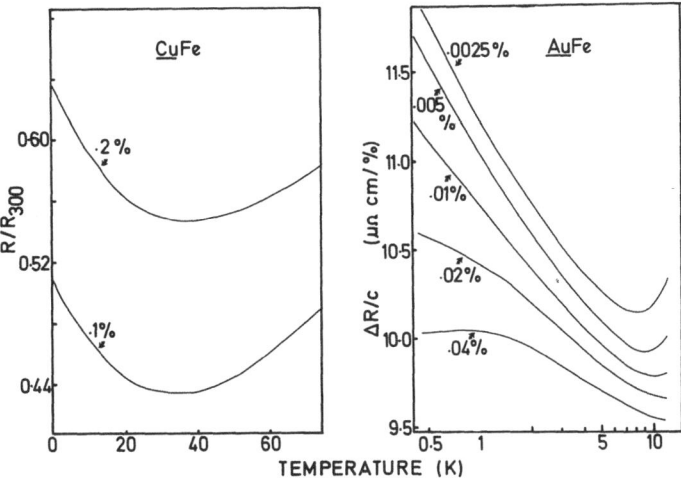

Figure 2. Two different representations of Kondo minima in the electrical
resistivity. Left, resistance divided by resistance at 300 K for
CuFe alloys from J.P. Franck et al. Proc. Roy. Soc. A262, 494
(1961). Right, the change in resistance per unit concentration
for AuFe alloys showing interaction effects at high concentrations
(Data from J.W. Loran et al. Phys. Rev. B2, 857 (1970)). For antiferromagnetic coupling, J between the conduction electrons and
the localized moment $T_k \approx E_F \exp{-1/Jn(E_F)}$ where $n(E_F)$ is the
density of states at the Fermi level. This relationship gives
very large variations of T_K for relatively small changes in J.

observed in condensed matter physics in the last two decades was that of
heavy-fermions. We plagiarize a paragraph from Fisk et al.[43] to introduce
the topic. The terms "heavy-electron metal" and "heavy-fermion system"
have been introduced to describe materials in which the electronic states
have a characteristic energy orders of magnitude smaller than in ordinary
metals. If we write the energy $\varepsilon(k)$ in the free-electron form
$[\varepsilon(k)=h^2 k^2/8\pi^2 m^*]$, then since the wave vectors k of the electron determined
by the interatomic spacing are not so different, the effective mass m^*
must be orders of magnitude larger than the free-electron value and in
some cases m^* is a fair fraction of the proton mass. These materials are
intermetallic compounds in which one of the constituents is a rare earth
or actinide atom, with partially filled 4f- or 5f-electron shells. At high
temperatures these materials behave as if the f-electrons were localized
on their atomic states, as in conventional rare earth and actinide compounds,
where any itinerant electrons are in states derived from loosely bound
atomic s-, p-, and d-orbitals. As the conventional materials are cooled,
the atomic orbitals due to the f-electrons order spontaneously, mostly
antiferromagnetically, less often ferromagnetically. By contrast, in the
heavy-electron systems some of the f-electrons become itinerant at low
temperatures and form a metallic state with characteristics described above.
(Jaap Franse describes some of the phenomena of ordering in heavy fermion
materials in his article in this section.) The concept that the f electrons
become itinerant at low temperatures is attractive in terms of low tem-

perature properties because it would give rise to a very high density of states at E_f. However it may not be correct. An alternative view is inherent in the Anderson impurity Hamiltonian. In this Hamiltonian, which is discussed in more detail in section B, C and D, one part derives from the s,p and d states and is rather free-electron-like. The other describes the spatially localized, and strongly correlated, f electrons. The f-levels are normally well below the Fermi level but cannot accept extra electrons because of the large Coulomb repulsion terms. Nevertheless, the f electrons do affect the ground state properties via the hybridization term in the Hamiltonian.

An alternative approach to low energy properties of narrow band systems is to probe directly the low energy excitations that control the thermodynamic properties. For a long time tunnelling spectroscopy was considered to be one of the most direct probes and it is still certain that it can contribute to this field (see the article by P. Wachter). However the article by L. Jansen et al. raises some serious doubts about the interpretation of the data from this method. It is sensible, even at this stage, to introduce one of the major results from high energy spectroscopy into the discussion. Thus figure 3 gives the energy positions of the lowest energy BIS and XPS peaks for the rare earths, from Lang et al.[44]. These energies correspond to those for the transitions between the lowest energy, Hunds rule, states f^n-f^{n+1} for BIS and f^n-f^{n-1} for XPS. n is the number of 4f electrons in the ground state and, with the exception of Eu and Yb, n is the position, N, of the element in the rare earth series. For Eu and Yb the special sta-

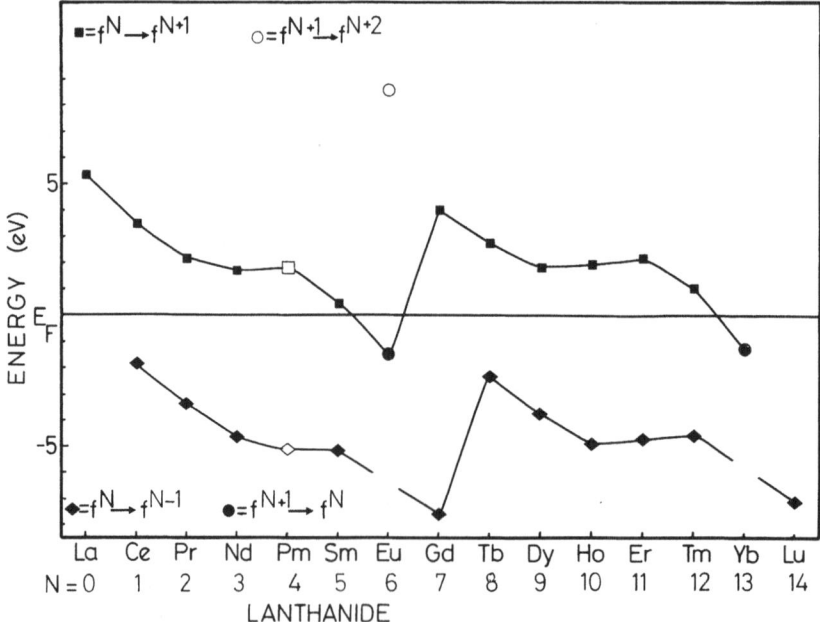

Figure 3. BIS and XPS binding energies of the f-states closest to E_F. (Note that for Ce we have omitted one peak near E_F which has been convincingly explained by Gunnarsson and Schönhammer[38] using a $1/N_f$ expansion. The data for this figure is taken from reference 44. Note also that here N refers to the number of the element in the rare earth series, so that for Eu and Yb N is not the number of f electrons in the ground state of the metal.

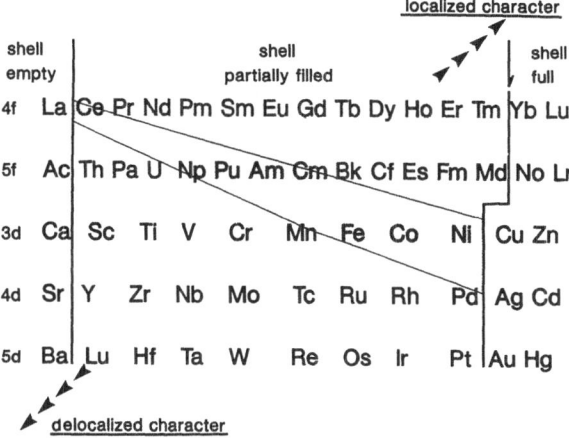

Figure 4. Quasi-periodic table indicating trends in the degree of localization of the f and d orbitals in rare earth actinides, and transition metals (after Smith and Kmetko[45]).

bility conferred on the Hund's rule ground states for the half filled and
filled shells, f^7 and f^{14}, leads to transfer of an electron from the conduction band to the 4f levels of the pure metals. The ground states of Eu
and Yb are thus f^{N+1} in the pure metals. Figure 3 immediately gives an
explanation for the interesting properties of Sm, Eu, Tm, and Yb. In each
case there is an XPS or BIS transition close to E_F, indicating that a small
chemical shift in the f level energy can lead to a change in f-count/valence
and even to mixed valence. For Ce the hybridization with the valence band
should be larger (see the discussion of figure 1) and when combined with
the special degeneracy effects in Ce, this can lead to anomalous effects
in the band states at E_F. We find it useful to bear Figure 3 in mind when
considering the chapters in the first section of this book.

Narrow band phenomena may be observed for atoms in compounds, if not
for the pure elements in most elements containing partially filled 4f, 5f,
3d, 4d and 5d electronic states. Figure 4 is a schematization of Smith and
Kmetko[45], which can be derived from two "rules of thumb". First, it is
generally recognized that the degree of localization of the 4f states
>5f>3d>4d>5d. i.e. the bandwidths of the states generally increase in this
order and the importance of atomic correlations decreases from the 4f to
the 5d states. However, there is a second rule of thumb concerning the
change in behaviour as one moves from left to right across the periodic
table. Then the expectation radius of the partially filled shell decreases,
and the importance of atomic correlations increases. The application of
the two thumb rules allows one to draw a broad band through the quasi-
periodic system of Smith and Kmetko in order to separate elements where
band structure effects dominate (bottom-left) from those where the bands
are narrow and atomic correlations dominate (top-right). Of course other
factors, such as general features of the chemical environment, interatomic
distances, orbital degeneracies, and the density of poorly correlated states
near the Fermi level, can all play a modifying role in this diagram. Thus
appropriate choices of these secondary parameters can often cause atoms
below the boundary region to act as if their electrons are localized and
correlated, or atoms above the boundary to act as if their electrons are
band-like. This form of materials tailoring is still in its infancy.

The book is presented in five major sections, with a short introduction and collection of discussion remarks for each section. The discussion remarks are not always attributed to their author because we felt that people could be hesitant to express useful but unproven opinions if they knew they were to be quoted. As noted elsewhere, we felt that the major new input to the field in recent years had been in spectroscopic methods and the theoretical methods to interpret them. However we did not restrict discussion to these points. In section A we group together papers and discussions of experimental studies of the ground state. Section B is given to theoretical descriptions. In section C we group papers on many-body descriptions and in section D papers on high energy spectroscopies and their interpretation. Finally in section E we group together discussions on the new high temperature super-conductors.

REFERENCES

1. N.F. Mott, Proc. Phys. Soc. London A62:416 (1949).
2. N.F. Mott, in: "Metal-Insulator Transitions", Taylor and Francis, London (1974).
3. J.A. Wilson, Advances in Physics 21:143 (1972).
4. J.B. Goodenough "Magnetism and the Chemical Bond", Interscience, New York (1963).
5. J. Hubbard, Proc. Roy. Soc. London, 276:238 (1963); 277:237 (1964); 281:404 (1964).
6. C. Herring in: "Magnetism" (eds G.T. Rado and H. Suhl), Academic Press, London (1966).
7. P.W. Anderson, Phys. Rev. 124:41 (1961).
8. J. Kondo, Prog. Theoret. Phys. 32:37 (1964);
    Solid State Physics, Vol. 23, ed. F. Seitz and D. Turnbull, Academic Press, New York, (1969).
9. See e.g. C. Kittel, Solid State Physics 22:1 (1968).
10. G.A. Sawatzky, W. Geertsma, and C. Haas, J. Mag. Mag. Mat. 3:37 (1976).
11. For a collection of articles on magnetic polarons, see J. Magn. Mag. Mater. 54-7:p1207ff (1985).
12. See e.g. Proceedings 1975 Discussion Meetings on magnetic Semiconductors, Ed. W. Zinn, N. Holland, Amsterdam 1976.
13. "Moment Formation in Solid", Ed. W. Buyers, Plenum, New York, 1984.
14. Valence Fluctuations 1985, Ed. E. Müller Hartmann, B. Roden and D. Wohlleben, J. Magn. Mag. Mater. 47-48:163 (1985).
15. C.M. Varma, Rev. Mod. Phys. 48:219 (1976).
16. See e.g. W. Meissner and B. Voigt, Ann. Phys. 7:761, 892 (1930); A.N. Gerritsen and J.O. Linde, Physica (Utrecht) 17:573 (1951); 18:877 (1951).
17. M. D. Daybell and W.A. Steyert, Rev. Mod. Phys. 40:380 (1968).
18. F. Steglich, J. Aerts, C.D. Bredl, W. Lieke, D. Meschede, W. Franz and H. Schäfer, Phys. Rev. Lett. 43:1892 (1972).
19. G.R. Stewart, Rev. Mod. Phys. 56:755 (1984).
20. N.F. Mott, Nobel prize Address, Rev. Mod. Phys. 50:203 (1978); P.W. Anderson, Nobel prize Address, Rev. Mod. Phys. 50:191 (1978).
21. P.A. Cox in:"The Electronic Structure and Chemistry of Solids", (Oxford Science Publ. Oxford, 1986).
22. N.W. Ashcroft and N.D. Mermin, p3 in: "Solid State Physics" (Publ. Holt, Rinehart and Winston, New York, 1976).
23. C. Kittel in: "Introduction to Solid State Physics" (6th edition) Wiley, New York, (1976).
24. E. Jensen and E.W. Plummer, Phys. Rev. Lett. 55:1912 (1985); E.W. Plummer, Physica Scripta T17:186 (1986) and references therein.

25. C.F. Fischer, "The Hartree-Fock Method for Atoms", J. Wiley, New York, (1977).
26. J.M. Fournier and L. Manes, in: "Structure and Bonding" 59/60:3,80 (1985).
27. J.P. Desclaux, Atomic and Nucl. Data tables 12:310 (1973).
28. D. van der Marel, Ph.D. thesis, Groningen University (1985).
29. "The Handbook of Chemistry and Physics", Ed. C. Weast, (publ. CRC press, Cleveland, 1975).
30. A.H. Wilson, Proc. Roy. Soc. A133:458 (1931); 134:277 (1931).
31. J.A. Wilson, p.215 in: "The Metallic and Non-Metallic States of Matter", ed. P.P. Edwards and C.N.R. Rao. Taylor and Francis (London 1985).
32. See e.g. J. Zaanen, G.A. Sawatzky and J.W. Allen, Phys. Rev. Lett. 55:418 (1985).
33. The concepts of "valence" and "valency", which still lead to some confusion, have interesting histories. The words themselves started to come into general use around 1870, thanks mostly to the work of Kekulé, (see, e.g., F.A. Kekulé, Z. Chem. 3:217 (1867); C.A. Russel, "The History of the Concept of Valency", (Leicester University Press, 1971)), although the ideas from which they were generated are usually traced back to Dalton's atomic theories and the law of constant, or rational, proportions. (J. Dalton, "A New System of Chemical Philosophy", Vol. 1, 1808, Cited by W.G. Palmer, "A History of the Concept of Valency to 1930", Cambridge University Press 1965). The period before 1870 was characterised by a great deal of confusion, although some interesting formulations of chemical structure and bonding were made. One of these was E. Frankland's (Phil. Trans. 142:440 (1852)) concept of an atom's "combining power" to explain the stoichiometry of inorganic and simple organic compounds.
All of these developments, of course, preceeded the development of an electronic theory of bonding. The present-day uses of the terms valence contain a mixture of the historical and the modern ideas. There are now four main usages, which can be a source of some confusion. Valency can be used in connection with the stoichiometry of a compound (i.e. the number of hydrogen atoms, or twice the number of oxygen atoms, which an atom of an element can replace, or with which it can combine). This is the use most closely related to the ideas of Kekulé. Secondly, valence is used in the general sense of chemical linking of atoms. Thirdly, there is a colloquial use to describe the actual links, or bonds in a molecule. Finally, there is a use, particularly among solid state physicists, meaning the number of electrons from an atom which contribute to the bonding (e.g. Sm and Yb are mixed valent in SmS_xSe_{1-x} and $YbAl_2$ respectively). This last usage has a superficial similarity to the atom combining power of Frankland.
34. See e.g. C.M. Varma, Rev. Mod. Phys. 48:219 (1976).
35. See e.g. F.A. Cotton and G. Wilkinson in: "Advanced Inorganic Chemistry", J. Wiley (1962).
36. D.R. Gustaffsson, J.D. McNutt and L.O. Roellig, Phys. Rev. 183:435 (1969).
37. B. Johansson, Philos. Mag. 30:469 (1974).
38. O. Gunnarsson and K. Schönhammer, Phys. Rev. Lett. 50:604 (1983), Phys. Rev. B28:4315 (1983); 31:4815 (1985).
39. J.W. Wilkins, Physics Today, 39:S22 (1986).
40. W. Meissner and B. Voigt, Ann. Phys. 7:761, 892 (1930).
41. J.P. Franck, F.D. Manchester and D.L. Martin, Proc. Roy. Soc. (London) A263:494 (1961).
42. R. M. Martin and J.W. Allen, J. Magn. Magn. Mater. 37-38:257 (1985).
43. Z. Fisk, H.R. Ott, T.M. Rice and J.L. Smith, Nature 320:124 (1986).
44. J.K. Lang, Y. Baer and P.A. Cox, J. Phys. F. 11:121 (1981).

45. J.L. Smith and E.A. Kmetko, J. Less. Common Metals, 90:83 (1983).
46. A.W. Overhauser, Adv. Phys. 27:243 (1978); Phys. Rev. Lett. 55:1916 (1985) and references therein.
47. W.A. Harrison, in: "Electronic Structure and the Properties of Solids", Freeman, San Francisco, (1980).
48. W.A. Harrison, Phys. Rev. B28:550 (1983).
49. H.J. de Boer and E.J.W. Verwey, Proc. Phys. Soc. A49:59 (1937).

A

GROUND STATE AND NEAR
GROUND STATE PHENOMENA

SECTION A: GROUND STATE AND NEAR GROUND STATE PHENOMENA

INTRODUCTION

The major impetus for research on materials displaying narrow band
phenomena has been, and will probably continue to be, their novel and interesting ground state properties, which have provided a challenge to
achieving a unified theory of the electronic structure of solids for at
least 50 years. The reason for this is that the properties of a particular
narrow band material are nearly always surprising, either from a theoretical
perspective or from a knowledge of empirical trends of wide band materials,
or even of other narrow band materials. The frequency with which the ad-
jective 'anomalous' is encountered in this field is a testimonial to the
role of surprises in the continuing love affair between condensed matter
researchers and narrow band materials.

One can identify classes of phenomena which have served as focal points
of researcher interest and in some cases have become trademarks of narrow
band behaviour. In no particle order of conceptual or historical importance,
these include:

- Occurrence of Mott-Hubbard insulators, insulating materials which
  have an odd number of valence electrons per unit cell and hence violate
  the Wilson rules.
- Occurrence of magnetically ordered solids.
- Occurrence of superconductivity near magnetic instabilities.
- Unusual electrical and thermal transport properties, such as a large
  T-linear specific heat coefficient, structure in the T-dependence of
  the resistivity, or a crossover with temperature variation between a
  Curie- and Pauli-type magnetic susceptibility.
- Magnetic moment effects in superconductors.
- Anomalous values of lattice constants, relative to trends.
- Complex magnetic and electrical phase diagrams with variation of tem-
  perature, pressure, or composition.

Two common models underlie most of the thinking by researchers, experimentalists and theorists alike. This is true to such an extent that it
is difficult to read even the experimental literature in this field without
knowing the elements of these models. One of them is the Hubbard Hamiltonian[1], which introduces an s-band, characterized by one-electron matrix
elements t_{ij} for electrons to hop from site i to site j, and an onsite

Coulomb repulsion U. When $U \gg t_{ij}$, the band is split, giving a Mott-Hubbard insulator for half filling. The other model is the impurity Anderson Hamiltonian[2] and its periodic generalization. In this model, a local s-orbital, again with on-site repulsion U, is hybridized with one-electron matrix element V to a continuum band characterized by width W or Fermi energy E_F. The local orbital binding energy relative to E_F is generally denoted ε_d or ε_f. For $U \gg \varepsilon_f > \rho V^2$, where ρ is the continuum density of states, there is integer occupation of the local orbital and a local moment is stable. The periodic version is sometimes regarded as an extended Hubbard model with two bands, one with $t_{ij} = 0$ and the other with $U = 0$, and hybridized to each other. Both models can be augmented to include orbital degeneracy in the bands, and off-diagonal site-Coulomb interactions among the narrow band electrons, leading to multiplet formation. In addition, the Anderson model can include a Coulomb repulsion between the local orbital and the continuum electrons, or even a nonzero U for the continuum electrons. There has been continuing debate as to how crucial all these additions are. The seriousness of this matter is shown by recent findings, discussed elsewhere in this volume, that including orbital degeneracy not only profoundly affects the physics of the impurity Anderson model, but also makes the solution of the model easy if the degeneracy is large enough.

The historical evolution of experimental work in this field is characterized by many intertangled threads and themes. From time to time one of these becomes emphasized when new techniques, materials or theories offer the promise of progress. Three recurring themes with a firm basis in observable phenomena are

1)   the transition from Mott-Hubbard insulator to metal,
2)   the quenching (or survival) of magnetic moments in solids, and
3)   the interplay between superconductivity and magnetism.

Note that these themes go beyond localized magnetism _per se_, in which the existence of atomic magnetic moments coupled by phenomenological exchange interactions is taken for granted, and the focus is on determining various types of magnetic order, and critical phenomena. There are other common themes, mentioned below, which seem to have a more philosophical flavour, and which are gradually being better defined as understanding improves. One such is whether to regard narrow band electrons as 'localized' or 'itinerant'.

With some important exceptions, transition metal systems were the most studied through the early 1970's. It was found that transition metal compounds exhibit a remarkable range of ground states, insulating, metallic, magnetic, and even superconducting, with apparently trivial, or modest changes in chemical composition, temperature or pressure. Much work was done to explore and systemize these behaviors, using mostly the Hubbard model as an underlying conceptual tool[3]. Several alloy systems were found to display rich phase diagrams as functions of composition, temperature and pressure. Two of these are $Ni(S_xSe_{1-x})_2$, where x varies between 0 and 1,[4] and $(V_{1-x}T_x)_2O_3$, where T is Cr or Ti and $x < 0.1$.[4] These systems have similar phase diagrams, with metallic, paramagnetic insulating and anti-ferromagnetic insulating regions. The latter system especially was characterized by very many different measurements and the metal-insulator transition was interpreted as a Mott transition, by which is meant the metal-insulator transition within the framework of the Hubbard model. The metallic regime was interpreted as a "nearly localized Fermi liquid", according to theoretical work[6] based on a variational wavefunction[7] for the ground state of the Hubbard model.

A non-transition metal system of great importance which was also studied in this period was cerium metal. By 1970, the initial finding[8] of the $\gamma \to \alpha$

transformation at about 8 kbar by room temperature compression and resistivity measurements, had been expanded by various workers to show finally[9] a complex phase diagram with five distinct solid phases. All five phases are metallic, but have differing superconducting and magnetic properties. The $\gamma \rightarrow \alpha$ transition is particularly interesting in that it is a first order transition involving loss of magnetic moments and a 15% decrease in volume, but no change in crystal structure. The first order phase transition line terminates in a critical point. By 1950 it had been proposed[10] from chemical arguments that this transition involved the promotion of the Ce 4f electron to the 5d state, and by 1970 the picture had been given substance in a calculation for a model[11] resembling the impurity Anderson model, but excluding hybridization, while including a conduction band-f electron Coulomb interaction. From this work there emerged the picture of a very sharp (approximately 10 meV) 4f state very near (approximately 100 meV) to the Fermi level in the γ phase, and moving from below it to above it in the $\gamma \rightarrow \alpha$ transition. This picture, which was very influential in determining thinking about other rare earth systems, has been challenged by theoretical and spectroscopic results of the 1970's and 1980's. Another description of the transition[12], based on Hartree-Fock solutions of the Anderson model, was also rather different from current thinking based on more recent treatments of this Hamiltonian.

Another category of transition metal research in this period was devoted to transition metal impurities in otherwise nonmagnetic materials, the underlying model generally being the impurity Anderson Hamiltonian. This work had three main aspects, the stability of the impurity's magnetic moment, the interaction of a stable local moment with conduction electrons in a normal metal, and the effect of a local moment on a superconducting host, especially in causing Cooper pair-breaking and a suppression of the host transition temperature T_c. For the second of these, the underlying model was the exchange Hamiltonian[13], or Kondo Hamiltonian, as it came to be called, in which the local moment has an exchange interaction of magnitude J with each conduction electron. This Hamiltonian can be derived[14] from the impurity Anderson model in the regime where local moments are stable, with the result that J is antiferromagnetic with magnitude J = $(2V^2U)/\varepsilon_d(\varepsilon_d+U)$. It has as a consequence the Kondo effect[15] in which, for antiferromagnetic coupling, the ground state of the interacting local moment-/conduction system is a singlet. Predictions of experimental consequences of the Kondo effect include a transition in the temperature dependence of the local moment magnetic susceptibility from Curie-like to Pauli-like, and a rise in the resistivity as the temperature is lowered through the Kondo temperature T_K. As with the transition metal compounds, much effort was expended in exploring the systematics of varying host and impurity. A focus of experimental effort was to test the Kondo picture and to determine T_K. A problem encountered was that of impurity-impurity interactions via the long range RKKY exchange interaction mediated by the conduction electrons[13], necessitating the study of very dilute systems. In superconductors, there was an effort to characterize the dependence on T_K of the rate of depression of the transition temperature T_c with impurity concentration. Much of this work is well summarized in review articles[16] written in the early 1970's.

In the very late 1960's and early 1970's there occurred a shift in emphasis from transition metal systems to rare earth systems, for which it was expected that the Anderson impurity model would be a better representation than for transition metal systems. With the exception of Ce, it had generally been assumed that rare earth 4f shells had integer occupations and possessed stable magnetic moments even in metallic environments. Thus there were studies of the Kondo effect and of T_c depression due to rare earth impurities. These also included cerium and it was found that with increasing pressure the pair-breaking strength first increases and then decreases[17]. From the decrease it was inferred that a loss of the cerium magnetic moment with pressure had occurred, a result consistent with the promotion model described above. Many cerium compounds, such as $CeRu_2$, were also found to display unusually small volumes, loss of magnetism, and occasionally superconductivity, and these were generally thought of as $4f^0$, i.e., tetravalent, cerium materials.

It was then very exciting to find experimental evidence that similar effects occur for other rare earth systems. It was observed that SmTe, SmSe, and SmS display pressure-induced semiconductor-metal transitions with a large volume decrease and loss of magnetic moments[18]. From evidence that all Sm sites are equivalent in both phases, and comparison of the atomic volumes of the two phases with the volumes of other rare earth (R) compounds, RTe, RSe, and RS, of known valence, it was inferred that the transitions occurred from divalent Sm to a state with non-integer occupation of the Sm 4f shell, between divalent $4f^6$ and trivalent $4f^5$. Analogous to the promotion model for cerium, it was thought that in the divalent phase the 4f level was separated from an empty 5d conduction band by a small gap, and that in the high pressure phase the gap had become negative to that electrons would leave the 4f level for the conduction band. This move-ment of a very narrow band, non-conducting because of a large U, into a broad band, is a metal-insulator transition mechanism distinguishable from that of the Mott transition. It results in a novel metallic situation with the Fermi level finally being pinned in the narrow 4f level. This situation came to be known as 'homogeneous mixed valence', and the terms 'valence fluctuations' or 'interconfiguration fluctuations' were introduced to charac-terize the idea that there was no energy barrier to electrons moving in and out of the 4f shell. It was also found that this situation obtains at atmospheric pressure in other compounds, such as SmB_6[19] and TmSe[20]. As with the transition metal systems, it is found that complex phase diagrams result from varying pressure, temperature and chemical composition. Many different measurements were made to characterize the electronic state of these materials, and an overview can be obtained from the proceedings of several international conferences on valence fluctuations, the first of which was held in 1976[21]. Neutron scattering has been very useful for study-ing the low energy spin dynamics associated with magnetic moment loss in the anomalous rare earth materials, including cerium ones.[22] The general picture described above for the Sm and Tm materials has seen only one major change in the time until the present, and that is the recognition that the mixed valent state may not always be metallic, but may sometimes have instead a very small gap[23]. The evidence for this is clearest in SmB_6 and TmSe. It seems certain that the theory of the gap is a subject where there is much room for progress. Since the late 1970's these materials have received less attention as the focus of interest has swirled on to other issues described next.

For cerium systems, a new view, which is approaching paradigm accep-tance, has arisen between 1980 and the present. It is that all metallic cerium materials lie in the Kondo regime of the impurity Anderson model, and can be at least partially characterized by a value of T_K. In this pic-ture, the α and γ phases of cerium have large (order of 1000 K) and small (order of 100 K) values of T_K, respectively, and a semiquantitatively suc-cessful 'Kondo volume collapse' model[24] of the phase transition can be made by ascribing a volume dependence to the coupling constant J, arising largely from the volume dependence of the hybridization V. In order for J and hence T_K to increase, the f-orbital occupation must decrease, i.e., the system becomes more mixed valent. Spectroscopy, combined with the modern theory of the impurity Anderson Hamiltonian, probably had the decisive effect in establishing this picture[25]. Nonetheless there was early thinking along this line; a proposal[26] of a high T_K for α-Ce, based on an analysis of the magnetic susceptibility; a proposal[27] that in the highest pressure phase of Ce there was as much f-occupation as for the γ-phase and influential theoretical arguments[28], confirmed later by spectroscopy, that ε_f for γ-Ce is of order 2 eV, an order of magnitude too large for the promotion model to work. Also, early controversies[29] on superconductivity depairing due to cerium impurities remarkably presaged the new developments. A theory of Kondo impurities in superconductors[30] in 1970-1971 predicted that de-pairing efficiency would display a maximum for T_K/T_c near 1. Thus it was possible[31] to interpret the pressure dependence of de-pairing due to Ce impurities, mentioned above, as being due to a continuous change of T_K for the cerium impurity from less than T_c to greater than T_c, rather than as a loss of cerium moment as in the promotion model. For those cerium materials which do not order magnetically at the lowest temperatures measured, or become superconductors, it is now common to use the term "Kondo lattice" materials, to convey the idea that the mechanism of the loss of magnetism is generically related to the impurity Kondo effect, although the theory of these lattice systems, and the competition between loss of magnetism and magnetic ordering due to the RKKY exchange interaction is still being developed. When T_K becomes comparable to electronic energies, e.g., exceeds a few thousand Kelvins, as may occur in the highest pressure, superconducting[32] phase of Ce, then the term mixed valent is more appropriate.

The late 1970's and early to mid 1980's also saw another focus of interest, and the birth of yet another name, the 'heavy Fermion' super-conductors. Actinide materials in general, and uranium materials in particular, which had been studied as narrow band materials for some time[33], began to receive much broader attention. The term heavy-Fermion is derived from the use of a very large effective mass to describe the unusually large T-linear specific heat-coefficients (γ-values) which occur in many rare earth and actinide materials, among them $CeAl_3$ and $CeCu_2Si_2$. Within the Kondo lattice picture, these large γ-values are explained by a small value of T_K, since the local moment spin entropy evolves over the temperature interval T_K, giving a γ of order $1/T_K$. Since the large γ is associated with spin fluctuations, the discovery[34] in 1979 that $CeCu_2Si_2$ is a super-conductor was an interesting surprise, followed somewhat later by the finding[35] that uranium compounds such as UPt_3 and UBe_{13} also are super-conductors with large normal state γ-values. Since the specific heat jump at T_c is comparable to γ, it appears that the heavy electrons are the super-conducting electrons. This idea is very important in the light of research done since 1976 on ternary and pseudoternary compounds of the form RMo_6S_8, RMo_6Se_8 and RRh_4B_4, with R = rare earth, remarkable materials where super-conductivity and ferromagnetism or antiferromagnetism coexist, but where it is believed that different electrons are involved in each type of or-dering[36]. Further, in some materials, e.g., UPt_3, many transport properties below T_c, such as ultrasound absorption[37], display a power law temperature dependence, suggesting a vanishing gap over portions of the Brillouin zone, anisotropic superconductivity, and the possibility of a non-phonon pairing mechanism. The observation[38] by neutron scattering of antiferromagnetic correlations on a very low energy scale in UPt_3 has given added credence to the possibility of an intimate connection between spin fluctuations and the superconductivity of this material. It has also been found that some uranium materials with large γ-values have magnetic or normal ground states, as with cerium systems. A large number of measurements of many kinds have been made for the heavy-Fermion systems[39], devoted to elucidating the elec-tronic state of the f-electrons, and the origin of the superconductivity. As yet, these have not resulted in anything approaching a complete or con-clusive picture, and the situation for uranium is especially open for lack of any unified interpretation of spectroscopic data. An overview can be obtained by perusing the proceedings of various recent international con-ferences[40].

Narrow band research also provides opportunities to better appreciate

19

and understand basic quantum mechanical ideas which are not new. One example is the continuing debate among narrow band researchers as to whether a 'localized' or 'itinerant' picture should be used to describe the narrow band electrons. The question is actually not well posed. In spite of strong experimental evidence from spectroscopy and from highly unusual transport properties that very large Coulomb interactions are at work to stabilize 'atomic' behaviour in these materials, the direct observation of heavy mass bands at the Fermi level by de Haas-van Alphen experiments[41] in, e.g., $CeSn_3$ and UPt_3, provides a very important reminder that the presence of translational symmetry has as a consequence quasiparticles at E_F for which the crystal momentum is a good quantum number.

There is a growing appreciation in the narrow band research community that this duality is compatible[42] with, and even required by, general ideas of many-body theory embodied in work which is rather old, e.g., the Fermi liquid theory and the Luttinger theorem[43]. Thus the near ground state properties have a Fermi liquid character over some scale of temperature and energy, e.g. T_K in a Kondo picture, while bare, 'undressed', atomic-like behaviour can be seen on larger energy scales. Elucidating the relation between the different energy scales is exciting and fundamental from both an experimental and theoretical viewpoint. The number and precise origin of the low energy scales for lattice systems remains unclear.

Another related example concerns the interpretation of time and energy scales in measurements on narrow band systems. For example, Mössbauer measurements in mixed valence systems show a single line with an isomer shift midway between that expected from either valence, while photoemission shows two sets of 4f spectra characteristic of the two valence states. It has often been said that the Mössbauer measurement is slower than the valence fluctuation rate, and hence averages the valence fluctuations, while the photoemission measurement, because it involves a high energy x-ray photon, is faster than the valence fluctuation rate and hence gives a 'snapshot' of the electronic state of the rare earth shell of the 4f electron. Underlying this description is the picture that the occupation of the 4f shell fluctuates in time, even at zero temperature, a classical view inconsistent with a quantum description of the ground state as stationary. Kohn and Lee[44] have carried through the exercise of a quantum analysis of the two experiments and shown that the essential features can be understood using only stationary states and paying proper attention to the possible final states of the electronic system for the measurement in question. In this analysis the period of the x-ray is irrelevant to the criterion for whether or not photoemission displays two lines. This analysis almost certainly has relevance for the questions posed by Wachter in this section.

The most recent development in the seemingly unending string of surprises in the narrow band materials is the finding of high temperature superconductivity in certain copper oxide materials. Later in this volume is a section devoted to this topic, and it suffices here to note that the mixed-valence community had begun to look back[45] to the transition metal compounds as long ago as 1984, and that there is already active research once again on the general subject of the Mott-Hubbard insulators and the Hubbard Hamiltonian. So the narrow band field has now come full circle in the time since the 1950's and 60's, and hopefully armed with new insights, understanding, and experimental and theoretical tools.

REFERENCES

1. J. Hubbard, Proc. Roy. Soc. (London) A276:238 (1963); Proc. Roy. Soc. (London) A277:237 (1964).
2. P.W. Anderson, Phys. Rev. 124:41 (1961).

3. J.B. Goodenough, in "Progress in Solid State Chemistry", Vol. 5, edited by H. Reiss (Pergamon, Oxford, 1972).
4. J.A. Wilson and D. Pitt, Philos. Mag. 23:1297 (1971).
5. D.B. McWhan, J.P. Remeika, T.M. Rice, W.F. Brinkman, J. Maita and A. Menth, Phys. Rev. Lett. 27:941 (1971);
    Phys. Rev. B5:2252 (1972).
6. W.F. Brinkman and T.M. Rice, Phys. Rev. B2:4302 (1970).
7. M. Gutzwiller, Phys. Rev. 134A:923 (1964);
    Phys. Rev. 137A:1726 (1965).
8. P.W. Bridgman, Proc. Am. Acad. Arts Sci. 62:207 (1927).
9. E. Franceschi and G.L. Olcese, Phys. Rev. Lett. 22:1299 (1969);
    D.B. McWhan, Phys. Rev. B1:2826 (1970);
    F.H. Ellinger and W.H. Zachariasen, Phys. Rev. Lett. 32:773 (1974).
10. W.H. Zachariasen, quoted by A.W. Lawson and T.Y. Tang, Phys. Rev. 76:301 (1949);
    L. Pauling, quoted by A.F. Schuck and J.H. Sturdivant, J. Chem. Phys. 18:145 (1950).
11. R. Ramirez and L.M. Falicov, Phys. Rev. B3:2425 (1971).
12. B. Coqblin and A. Blandin, Adv. Phys. 17:281 (1968).
13. C. Zener, Phys. Rev. 81:440 (1951);
    M.A. Ruderman and C. Kittel, Phys. Rev. 96:99 (1954); T. Kasuya, Progr. Theor. Phys. 16:45 (1956);
    K. Yosida, Phys. Rev. 106:893 (1957).
14. J.R. Schrieffer and P.A. Wolff, Phys. Rev. 149:491 (1966).
15. J. Kondo, Progr. Theor. Phys. 32:37 (1964).
16. See "Magnetism", Vol. 5, edited by H. Suhl, (Academic, New York, 1973).
17. M.B. Maple, J. Wittig, and K.S. Kim, Phys. Rev. Lett. 23:1375 (1969).
18. A. Jayaraman, V. narayanamurti, E. Bucher, and R.G. Maines, Phys. Rev. Lett. 25:368 (1970);
    Phys. Rev. Lett. 25:1430 (1970);
    M.B. Maple and D. Wohlleben, Phys. Rev. Lett. 27:511 (1971).
19. E.E. Vainshtein, S.M. blokhin, and Yu. B. Paderno, Fiz. Tverd. Tela 6:2909 (1964) [Sov. Phys. Solid State 6:2318 (1965)];
    R.L. Cohen, M. Eibschutz, and K.W. West, Phys. Rev. Lett. 24:383 (1970);
    J.C. Nickerson, R.M. White, K.N. Lee, R. Bachman, T.H. Geballe and G.W. Hull, Jr., Phys. Rev. B3:2030 (1971).
20. See E. Bucher, K. Andres, F.J. diSalvo, J.P. Maita, A.C. Gossard, A.S. Cooper and G.W. Hull, Phys. Rev. B11:500 (1975) for a compilation of work on Tm monochalcogenides.
21. "Valence Instabilities and Related Narrow-Band Phenomena", edited by R.D. Parks (Plenum, New York, 1977);
    "Valence Fluctuations in Solids", edited by L.M. Falicov, W. Hanke, and M.B. Maple, (North-Holland, Amsterdam, 1981);
    Proceedings of the 4th International Conference on Valence Fluctuations, Cologne, 1984, edited by E. Müller-Hartmann, B. Roden and D. Wohlleben, (North-Holland, Amsterdam, 1985);
    Proceedings of the 5th International Conference on Crystalline Field and Anomalous Mixing Effects in f-electron Systems, Sendai 1985, edited by T. Kasuya, (North-Holland, Amsterdam, 1985).
22. See, e.g., section 2.4.4 of "Valence Fluctuation Phenomena", J.M. Lawrence, P.S. Riseborough and R.D. parks, Reports on Progress in Physics 44:1-84 L(1981).
23. N.F. Mott, Philos. Mag. 30:403 (1973);
    J.W. Allen, B. Batlog and P. Wachter, Phys. Rev. B20:4807;
    R.M. Martin and J.W. Allen, in: "Valence Fluctuations in Solids", edited by L.M. Falicov, W. Hanke and M.B. Maple (North-Holland, Amsterdam, 1981) p.85.
24. J.W. Allen and R.M. Martin, Phys. Rev. Lett. 49:1106 (1982);
    M. Lavagna, C. Lacroix and M. Cyrot, Phys. Lett. 90:210 (1982);
    R.M. Martin and J.W. Allen, J. Magn. and Magn. Mater. 47&48:257 (1985).
25. J.W. Allen, S.-J. Oh, O. Gunnarsson, K. Schönhammer, M.B. Maple, M.S.

Torikachvili, and I. Lindau, Adv. in Phys. 35:275 (1986).

26. A.S. Edelstein, Phys. Rev. Lett. 20:1348 (1968).

27. H.H. Hill and E.A. Kmetko, J. Phys. F: Metal Phys. 5:1119 (1975).

28. B. Johansson, Philos. Mag. 30:469 (1974).

29. M.B. Maple, Appl. Phys. 9:179 (1976).

30. E. Müller-Hartmann and J. Zittartz, Z. Physik 234:58 (1970).

31. W. Gey and E. Umlauf, Z. Physik. 242:241 (1971).

32. J. Wittig, Phys. Rev. Lett. 21:1250 (1968).

33. See, e.g., M.B. Brodsky, in "Valence Instabilities and Related Narrow Band Phenomena", edited by R.D. Parks, (Plenum, New York, 1977), p. 351.

34. F. Steglich, J. Aerts, C.D. Bredl, W. Lieke, D. Meschede, W. Franz, and H. Schäfer, Phys. Rev. Lett. 43:1892 (1979).

35. H.R. Ott, H. Rudiger, Z. Fisk and J.L. Smith, Phys. Rev. Lett. 50:1595 (1983);
G.R. Stewart, Z. Fisk, J.O. Willis and J.L. Smith, Phys. Rev. Lett. 52:679 (1984).

36. M.B. Maple, in "Advances in Superconductivity", edited by B. Deaver and John Ruvalds, NATO ASI Series, Series B:Physics, Vol. 100, (Plenum, New York, 1983) p.279.

37. D.J. Bishop, C.M. Varma, B. Batlogg, E. Bucher, Z. Fisk and J.L. Smith, Phys. Rev. Lett. 53:1009 (1984).

38. G. Aeppli, A. Goldman, G. Shirane, E. Bucher, and M.-Ch. Lux-Steiner, Phys. Rev. Lett. 58:808 (1987).

39. G.R. Stewart, Rev. Mod. Phys. 56:755 (1984).

40. Proceedings of the International Conference on Anomalous Rare Earths and Actinides, Grenoble, 1986, edited by J.X. Boucherle, J. Flouquet, C. Lacroix, and J. Rossat-Mignod, (North-Holland, Amsterdam, 1987); Proceedings of the 5th International Conference on Valence Fluctuations, Bangalore, 1987, edited by L.C. Gupta and S.K. Malik, (Plenum, New York, 1987).

41. W.R. Johanson, G.W. Crabtree, A.S. Edelstein and O.D. McMasters, Phys. Rev. Lett. 46:504 (1981);
J. Appl. Phys. 52:2134 (1981);
L. Taillefer, R. Newbury, G.G. Lonzarich, Z. Fisk and J.L. Smith, J. Magn. and Magn. Mat. 63&64:372 (1987).

42. R.M. Martin, Phys. Rev. Lett. 48:362 (1982);
N. d'Ambrumenil and P. Fulde, in Proceedings of the International Conference on Valence Fluctuations, Cologne, 1984, edited by E. Müller-Hartman, B. Roden and D. Wohlleben (North-Holland, Amsterdam, 1985), p. 1.

43. J.M. Luttinger, Phys. Rev. 119:1153 (1960).

44. W. Kohn and T.K. Lee, Philos. Mag. A45:313 (1982).

45. J.W. Allen, in Proceedings of the 4th International Conference on Valence Fluctuations, edited by E. Müller-Hartmann, B. Roden and D. Wohlleben (North-Holland, Amsterdam, 1985) p.168.

FROM TRANSITION METALS TO NORMAL RARE EARTH METALS VIA HEAVY FERMION SYSTEMS

D. M. Edwards

Department of Mathematics

Imperial College, London SW7 2BZ, U.K.

In these brief notes I shall use the simplest model of an Anderson lattice with an orbitally non-degenerate d or f level as a reference system:

$$H = \sum_{k\sigma} \varepsilon_k\, c^+_{k\sigma}\, c_{k\sigma} + V \sum_{k\sigma} (c^+_{k\sigma} f_{k\sigma} + f^+_{k\sigma} c_{k\sigma}) + E_f \sum_{i\sigma} n_{i\sigma} + U \sum_i n_{i\uparrow} n_{i\downarrow} .$$

(1)

Clearly in various regimes orbital degeneracy, spin-orbit coupling, realistic $\underline{k}$-dependent hybridization, direct f-f hopping, direct c-f exchange, multiple conduction bands, etc. will be more or less important. To obtain a metal, even in the presence of a possible hybridization gap, we assume a non-integral number of electrons. This may be achieved by the formal device of introducing a second, unhybridized, conduction band which acts merely as an electron reservoir.

1. Band v. Many-body;

There need be no conflict between band calculations which place the f level just above the Fermi level and many-body theories in which the bare f level E_f is one or two eV below the Fermi level. The relationship between the two approaches may be seen as follows.

The exact form of the one-particle Green function is

$$G_\sigma = \begin{bmatrix} G^\sigma_{ff} & G^\sigma_{fc} \\ G^\sigma_{cf} & G^\sigma_{cc} \end{bmatrix} = \begin{bmatrix} E - E_f - \Sigma_\sigma & - V \\ - V & E - \varepsilon_k \end{bmatrix}^{-1}$$

(2)

where $\Sigma_\sigma(\underline{k},E)$ is a proper self-energy. Assuming the E dependence of Σ to be dominant the Dyson equation is thus

$$(E - E_f - \Sigma_\sigma(E)) (E - \varepsilon_{\underline{k}}) = V^2 .$$

(3)

Assuming isotropic bands the Fermi wave-vector $k_{F\sigma}$ is determined by the total number of electrons of spin σ, via the Luttinger theorem, so that at the Fermi level (E = 0) we have $E_f + \Sigma_\sigma(0) = V^2/\varepsilon_\sigma$ where $\varepsilon_\sigma = \varepsilon_{k_{F\sigma}}$. A band calculation for given electron number necessarily gives the correct Fermi surface volume and the effective f potential $E^{eff}_{f\sigma}$ is essentially energy-independent. It must therefore be given by V^2/ε_σ and the energy bands satisfy

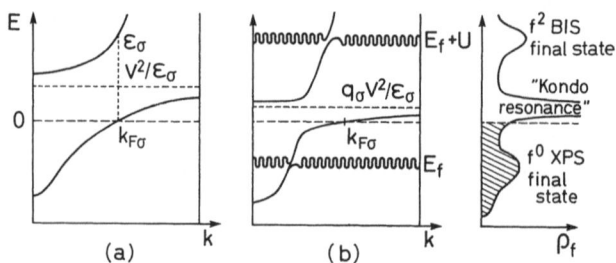

Figure 1. For explanation, see text

$$(E - V^2/\varepsilon_\sigma)(E - \varepsilon_{\underline{k}}) = V^2 . \tag{4}$$

as shown in fig. 1(a). The "true" spectrum satisfies (3) and, near E = 0, $E_f + \Sigma_\sigma(E) \simeq V^2/\varepsilon_\sigma + (d\Sigma_\sigma/dE)_0 E$. Hence near E = 0 the quasiparticle bands satisfy

$$(E - q_\sigma V^2/\varepsilon_\sigma)(E - \varepsilon_k) = q_\sigma V^2 \tag{5}$$

where $q_\sigma = \{1 - (d\Sigma_\sigma/dE)_0\}^{-1}$. Comparing (4) and (5) it is easily seen that the ratio of quasiparticle mass to band mass is given by $m^*/m = q_\sigma^{-1}$ and the ratio of quasiparticle mass to conduction band mass is $m^*/m_0 = q_\sigma^{-1}(\varepsilon_\sigma/V)^2$. Furthermore the f weight in the quasi-particle band at the Fermi level, and hence the Migdal discontinuity in $<f_{k\sigma}^+ f_{k\sigma}>$, is equal to q_σ. For nearly localized f electrons (U large and $1 - n_f << 1$, $n_f = <n_\uparrow> + <n_\downarrow>$), $q_\sigma << 1$. Then most of the f weight is far from the Fermi level and its distribution is indicated roughly by the atomic limit

$$G_{ff}^\sigma \Big|_{V=0} = \frac{1 - <n_{-\sigma}>}{E - E_f} + \frac{<n_{-\sigma}>}{E - E_f - U} \simeq \frac{<n_\sigma>}{E - E_f} + \frac{1 - <n_\sigma>}{E - E_f - U} . \tag{6}$$

The "true" spectrum and f density of states which corresponds to the above considerations is sketched in fig. 1(b), for comparison with the band calculation picture of fig. 1(a). The band picture is valid within its own ground-state framework, being capable of yielding essentially correct electron densities $\rho_\sigma(\underline{r})$, ground-state energy and a reasonable Fermi surface. However it is not designed to give the "true" excitation spectrum which includes structure around E_f and $E_f + U$, seen in XPS and BIS respectively, as well as strongly narrowed structure just above the Fermi level. These narrowed quasiparticle bands correspond to the "Kondo resonance" of the impurity model.

Undoubtedly the most important role of a band calculation is to determine the complex Fermi surfaces of real materials and thus interpret dHvA measurements. The spin density functional method is reasonable for most 3d systems, and some actinide and rare-earth systems without magnetic order, but cannot handle situations where orbital symmetry is broken, e.g. magnetically-ordered Ce systems where the unrestricted Hartree-Fock method[1] has been applied or cases where crystal field splitting is greater than the width of the narrowed f resonance. The latter case in $CeCu_2Si_2$ has been treated by d'Abrumenil and Fulde[2] and clearly leads to a Fermi surface which differs from that which would be obtained in a straight-forward density functional calculation.

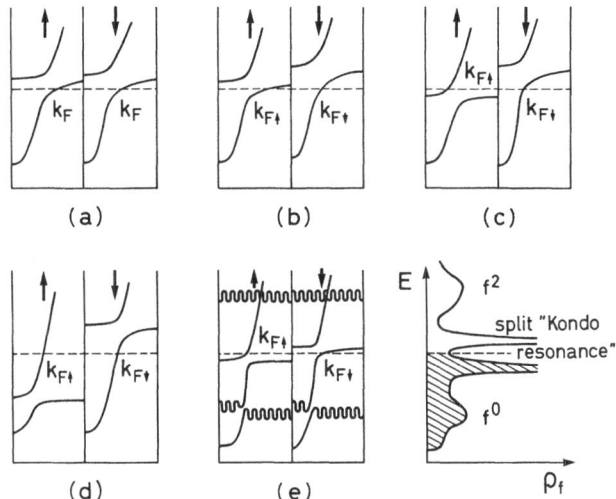

Figure 2. For explanation, see text.

2. The ideal theory: This would handle correlation effects in systems
ranging from itinerant 3d electron systems, e.g. Ni, to localized 4f systems
such as Gd. The nearly localized intermediate valent and heavy fermion
systems lie in between and are flanked by systems such as MnSi and Pr, both
with $m^*/m \sim 4$, on the itinerant and localized sides respectively.
Specialized theories such as those developed for impurity models, e.g. Bethe
ansatz, 1/N expansions, lack the necessary range and the most likely,
although difficult, approach is the standard many-body perturbation
expansion in V and U which leads to (2). The self-energy diagrams are
identical with those of the Hubbard Model but with modified propagators
$[E - E_f - V^2/(E - \varepsilon_k)]^{-1}$. This approach underlies the spin fluctuation
theories of Moriya and Kawabata [3] and Lonzarich [4] for strongly paramagnetic
and weakly ferromagnetic itinerant electron systems. However the existing
microscopic theory cannot handle the large U limit and it is a strength of
Lonzarich's more phenomenological treatment that all many-body renormal-
isations are absorbed in a few disposable parameters. The microscopic
theory for large U is simpler in the strongly magnetic case where I have
adapted previous work on the Hubbard model to the Anderson lattice, as
reported at the Bangalore conference [5].

3: Transition to magnetism: In fig. 2 it is shown how a transition may take
place from a paramagnetic heavy fermion system to a standard rare-earth metal
with magnetically-ordered localized f electrons. For simplicity ferromagnetic
order is assumed. Figs. 2(a) - (d) show band structure pictures (c.f. fig
1(a)) and for each such picture we can draw a "true" spectrum e.g. fig. 2(e)
corresponds to fig. 2(c). We see (a) a paramagnet with magnetization M = 0
(b) a weak magnet (M≠0) (c) a nearly strong ferromagnet (M≠0) (d) a standard

rare earth metal ($M=M_{max}$). As M increases we switch from "itinerant magnetism" ($k_{F\uparrow} > k_{F\downarrow}$) to "rare earth magnetism" with a negatively polarized conduction band ($k_{F\uparrow} < k_{F\downarrow}$). Somewhere between (b) and (c), at M_{crit} say, the volume of the ↑ spin Fermi surface changes suddenly by one electron. In (a) and (b) the total volume within the Fermi surfaces corresponds to $n_c + n_f$ electrons (conduction and f electrons itinerant) and in (c) and (d) it corresponds to $n_c + n_f - 1 \approx n_c$ electrons. Clearly from (e) the "Kondo resonance" is split in a ferromagnet, the splitting being on the same energy scale as its width. For $M < M_{crit}$ both ↑ and ↓ spin resonances lie above the Fermi level; for $M > M_{crit}$ the ↑ spin resonance moves below. In ref. 5 it is shown how the breakdown of magnetic order which occurs from (d) to (a) begins to occur as V is increased. It is found there that magnetic instability occurs when $E_0 < W \exp(-2W|E_f|/V^2)$ where E_0 is a typical magnon energy and 2W is the width of the conduction band. As the instability is approached there is rapid increase of mass enhancement m^*/m in the ↓ spin band. It will be interesting to study changes in Fermi surface and masses by dHvA in a pressure-driven magnetic phase transition.

4. Ultimate aim: (i) Use a suitable method of band calculation to obtain a Fermi surface in agreement with dHvA, if available. (ii) Fit parameters of a generalised Anderson lattice (degenerate f level, multiple conduction bands) to the band structure; use spectroscopic data to determine E_f and U. (iii) Use perturbation theory in U to develop a self-consistently renormalized spin fluctuation theory (generalised to include spin-orbit coupling and valid for large U). (iv) Hence calculate everything of interest via one- and two-particle Green functions at zero and finite temperature. This includes magnetic, quadrupolar and superconducting phase transitions.

References

1. O. Sakai, M. Takeshige, H. Harima, K. Otaki and T. Kasuya, J. Magn. Magn. Mat. 52 18 (1985).
2. N. d'Abrumenil and P. Fulde, J. Magn. Magn. Mat. 47-48 1 (1985).
3. e.g. A. Kawabata, J. Phys. F 4 1477 (1974).
4. G. Lonzarich, J. Magn. Magn. Mat. 54-57 612 (1986).
5. D. M. Edwards, Proc. Fifth International Conf. on Valence Fluctuations, Bangalore (Plenum 1987).

QUASIPARTICLES IN NARROW BAND SYSTEMS

Peter Fulde

Max-Planck-Institut für Festkörperforschung
7000 Stuttgart 80
Fed. Republic of Germany

When electrons are moving through a molecule or solid, two distinct features influence their motion. One is the kinetic energy gain due to the overlap of orbitals centered on different atoms. In a solid it is characterized by the band width W and results in charge fluctuations at the different atomic sites. The other one is the Coulomb repulsion between the electrons. It is minimized when the electrons stay apart as far as possible. Clearly, the kinetic energy gain and the Coulomb repulsions are competing effects, as the latter try to reduce charge fluctuations to a minimum. More precisely it is the ratio (U-K)/W which governs the size of the charge fluctuations, where U is the repulsion between two electrons at the same site and K is the one between two electrons on neighboring sites.

When $(U-K) \lesssim 1$, as it is the case when s and p electrons and often d electrons are considered, one can build electron correlations into the theory by using uncorrelated electrons as a starting point. Even in those comparatively weakly correlated systems the effects of correlations on the ground state and excited state energies are considerable (examples are semiconductors or ordinary metals).

In narrow band systems $(U-K)/W > 1$. Those systems are so difficult to treat because the uncorrelated state would be an extremely poor starting position. The question arises, whether it is nevertheless possible to determine the energy dispersion $E(\underline{k})$ of a quasiparticle, which has been added to the ground state of the system. We consider deliberately only one quasiparticle, because if there are many of them they influence also the $\underline{k}$ dependence of $E(\underline{k})$ as it is known from the work of Landau (see e.g. [1]).

The traditional way of calculating $E(\underline{k})$ is by solving a single-particle Schrödinger equation for which sophisticated methods have been developed by Andersen [2] and others. This should hold true also for a narrow band (or strongly correlated) system. The only question is, what is the potential which enters into such an equation? For this potential usually the local-density approximation (LDA) is made, i.e. for the exchange and correlation contribution to this potential a form is used which is derived from the homogeneous electron gas. Despite of its great success it is clear that this approximation has its limitations. This is so because e.g. not all forms of electron correlation effects which may be present in a solid do also occur in a homogeneous electron gas. Also in narrow band systems like the heavy fermion systems, there are correlations present which do not occur in the homogeneous electron gas. They are related to the formation of singlets between f electrons of e.g. rare earth ions and electrons of the ligands. This singlet formation can be studied by considering the variational wave function of Varma and Yafet [3] or even simpler, by considering a molecular analogue of it [4]. Expressed more formally, the LDA replaces

$$E(\underline{k}) = \frac{k^2}{2m} + \sum (\underline{k}, E(\underline{k})),$$
(1)

where $\sum(\underline{k},\omega)$ is the electron self-energy, by [5]

$$E(\underline{k}) = \frac{k^2}{2m} + \sum (k_F, E(k_F))$$
(2)

Here k_F is the Fermi momentum. Due to the singlet formations $\sum (\underline{k},\omega)$ is strongly ω dependent in the vicinity of $\omega = E(k_F)$. Therefore the replacement of Eq. (1) by (2) is not possible and the LDA cannot be used. Are we then completely stuck? The answer is no. It turns out that one can cast the difficulties related to the many-body or correlation problem into the form of a small number of parameters which must be determined from experiments. In the case of heavy fermion Ce compounds such as $CeCu_2Si_2$ or $CeAl_3$, it is just one parameter which has to be introduced. The way this is achieved is as follows. Let us characterize the potential which the electrons experience, in terms of energy dependent phase shifts $\delta_l^A(\omega)$ where A refers to the different atoms and l to the angular momentum of the in- or outgoing wave. The singlet formations result in low lying excitations of predominantly f character. Therefore it is reasonable to assume that the $\delta_{l=3}^{Ce}(\omega)$ phase shifts must be modified as compared with the ones which follow from LDA, but that for the remaining phase shifts the LDA may be used. The phase shifts $\delta_{l=3}^{Ce}(\omega)$ are, of course, unknown. However, one knows that for e.g. Ce^{3+} the $4f^1$ configuration is characterized by J=5/2 (Hund's rule). The crystal field (CEF) ground state is a Kramer's doublet of symmetry Γ_0 which can be determined from neutron scattering experiments. The CEF excited states are high up in energy. Only for an incoming wave of Γ_0 symmetry do we then have to introduce a phase shift, because otherwise the f charge distribution would differ from that which follows from the neutron data. This phase shift is expanded around E_F, i.e.

$$\delta_{\Gamma_0}^{Ce} (\omega) = \delta(E_F) + \frac{1}{k_B T^*} (\omega - E_F)$$
(3)

which defines two parameters $\delta(\varepsilon_F)$ and T^*. By using Eq. (3) and determining all other phase shifts with the help of the LDA one obtains E(k) for a given Ce compound. It turns out, and indeed is suggestive, that $\delta(E_F)$ is determined by a self-consistency requirement. Namely $\delta(E_F)$ determines the f electron count which enters into the potential of the other electrons and therefore influences E_F. Therefore the theory contains only T^* as a parameter, which is determined by adjusting the calculated density of states e.g. to the specific heat coefficient γ:

Reducing the strong correlation problem to a single particle problem with unconventional, i.e. strongly renormalized potential parameters corresponds to a mean field approximation [6]. The mean field refers thereby to an auxiliary boson which describes the strong correlations.

A description of the details of the above computation scheme can be found in Refs. [7,4]. Results for different compounds are given in Refs. [7,8]. It should be mentioned that the present introduction of the phase shift (3) differs from the one given by Nozières [9]. In his case of a single ion, the starting point is an external spin coupled through a Kondo Hamiltonian to the conduction electrons. The screening of the spin by the conduction electrons is accomplished through a proper phase shift of the latter. In our case the phase shift (3) generates the f electron charge distribution and does not screen an f electron moment.

Using the LDA for the other scattering channels is of course also apt for improvements. For example, it is well known from Pr metal, that the higher CEF levels of the 4f shell affect the effective mass of the conduction electrons [10]. The latter are dressed by virtual CEF excitations to which they couple, similarly as they are dressed by a virtual phonon cloud. The effect results again in a ω dependence of $\sum(k,\omega)$ near E_F and is not contained in a LDA description. Finally we want to mention briefly what happens as the temperature T increases. For T >> T*, but much less than the energy scales, over which the conduction electron self-energy varies, a "new" Fermi surface establishes itself as compared with the one at T=0. While at T=0 the Fermi surface has predominantly f electron character, at T >> T* the f electrons can be treated as being localized, or like core states (e.g. like in Pr). Then the Fermi surface is determined by the conduction electrons, which interact with the f electrons through generalized s-f exchange and Coulomb interactions.

REFERENCES

1. D. Pines and P. Nozières, "The Theory of Quantum Liquids", Vol. I (Benjamin Publ., New York, 1966)
2. O.K. Andersen, Phys. Rev. B 12, 3060 (1975)
3. C.M. Varma and Y. Yafet, Phys. Rev. B 13, 2950 (1976)
4. P. Fulde, J. Keller and G. Zwicknagl, to appear in "Solid State Physics" edit. by H. Ehrenreich and D. Turnbull
5. U. von Barth, Phys. Rev. A 20, 1693 (1979)
6. see e.g. P. Coleman, Phys. Rev. 29, 3035 (1984)
7. N.d'Ambrumenil and P. Fulde, J. Magn. Magn. Mat. 47+48, 1 (1985)
8. J. Sticht, N. d'Ambrumenil and J. Kübler, Z. Phys. B 65, 149 (1986)
9. P. Nozières, J. Low Temp. Phys. 17, 31 (1974)
10. R.M. White and P. Fulde, Phys. Rev. Lett. 47, 1540 (1981)

DYNAMICAL MAGNETIC FLUCTUATIONS IN NARROW BAND METALS

Piers Coleman[a] and G.G. Lonzarich[b]

[a]Serin Physics Laboratory, Rutgers University, P.O. Box 849
Piscataway, NJ 08855, U.S.A.

[b]Cavendish Laboratory, University of Cambridge
Madingley Road, Cambridge, CB3 0HE, U.K.

A growing body of evidence suggests that narrow band d metals and f metals might be understood within a single unified picture. Behaviour characteristic of heavy fermion f systems has been observed in a variety of d metals on the verge of a magnetic phase transition or a Mott transition at low temperatures[1,2], whilst recently via the de Haas van Alphen effect[3,4] and through considerations of other low temperature phenomena[5-7] it has been demonstrated that the f electrons in heavy fermion rare earth and actinide metals can be described in terms of an itinerant electron picture.

However, despite the similarities between these kinds of systems, there is surprisingly little overlap between the theoretical approaches so far taken to describe them. Traditionally an itinerant picture, normally based on the one band Hubbard model, is used to describe d metals, whereas for rare earth and actinide systems with their more localised f-shells, a starting picture of local magnetic ions is favoured and the Anderson lattice model is adopted. In most analyses presented so far an apparently different emphasis has been given by the two lines of theoretical development on the role played by *dynamical magnetic fluctuations*.

In conventional analyses based on the Hubbard model overdamped spin fluctuations with a strongly wavevector dependent relaxation spectrum, or paramagnons, are of major importance and give rise to a renormalisation of the quasiparticle energy spectrum and to spin dependent quasiparticle interactions[8-17]. In standard treatments of the Anderson lattice model, based on leading order expansions in the inverse of the degeneracy $N = 2j+1$ of the atomic f levels, magnetic fluctuations are *not* as manifestly important. (Discussions of this model and related problems may be found, for example, in references 18 to 28). It seems likely, however, that magnetic fluctuations also play a major part in the latter model when it is treated to a sufficiently high order in 1/N to describe real systems.

As a first step in broadening our understanding of the role of slow fluctuations of the magnetisation we consider a phenomenological quantum Ginzburg Landau description. The partition function is assumed to be given by a sum of exp(-βF[M]) over functions M(r,τ) of position r and time τ in volume V and in [0,ħβ =ħ/k_BT], respectively, where F[M] is a free energy functional of M[13,16,17,29,30]. For the purpose of illustration we treat the magnetisation as a scalar field in a system possessing translational and rotational invariance and *assume* that F[M] can be expanded in a power series to fourth order in M as

$$\hbar\beta F[M] = \hbar\beta F[0] + \frac{1}{2}\int d1d2M(1)a(1-2)M(2) + \frac{b}{4}\int d1M(1)^4 + ... \qquad (1)$$

where $1 = (r_1,\tau_1)$, the integrals are over all r in volume V and τ in [0,ħβ], b is the (local) mode coupling parameter, and a(r,τ) is

$$a(r,\tau) = \frac{1}{\hbar\beta V} \sum_{\substack{q<q_c \\ |\omega_n|<\omega_c}} a(q,i\omega_n) e^{i(q.r-\omega_n\tau)} \qquad (2)$$

$$\omega_n = 2\pi n/\hbar\beta, \quad n = 0, \pm1, \pm2, ...,$$

where a(q,ω) is the inverse of the wavevector and frequency dependent magnetic susceptibility in the *absence* of mode coupling, and q_c and ω_c are suitably chosen cut-off parameters in the Fourier series defining a(r,τ) and M(r,τ). For low frequencies we further assume that we may write a(q,ω) in the form

$$a(q,\omega) = a(q) - \frac{i\omega}{u(q)} + ..., \qquad (3)$$

where a(q) = a(q,ω=0) and u(q) is a q dependent parameter associated with the long time relaxation rate of magnetic fluctuations of wavevector q. If we neglect the mode coupling term in equation (1) and terms beyond first order in ω in equation (3) the relaxation rate is given as u(q)a(q).

At low q, a(q) may be expanded as

$$a(q) = a + cq^2 + ..., \qquad (4)$$

and, in the collisionless quasiparticle regime, we may also write

$$u(\mathbf{q}) = \gamma q + \dots,$$

(5)

where a, c and γ are constants (see e.g. ref. 31). To this order the Ginzburg Landau functional is characterised by four primary parameters, namely a, b, c and γ, the first three of which also appear in an equivalent classical Ginzburg Landau free energy functional in which the time dependence of the fluctuations is neglected. Recently it has been shown that a model having the above general structure, based on four principal ground state parameters, on a vector (rather than a scalar) field, and on a self-consistent Hartree approximation for the mode coupling, provides a surprisingly accurate quantitative description of the leading temperature dependences of the magnetic equation of state[2,31] and of the heat capacity[2,32] in a variety of transition metals on the brink of ferromagnetic long range order at low temperatures, in which only relatively long wavelength magnetic fluctuations play a major role.

In the case of nearly localised f electrons, magnetic fluctuations over all of q-space are important and an approximate description in the above quantum Ginzburg Landau model, extended to take account of the underlying lattice and the strong amplitude of the fluctuations, would in lowest order effectively take $a(\mathbf{q})$ and $u(\mathbf{q})$ to be constants independent of q over wide regions in the Brillouin zone. As was found for the transition metals, a phenomenological description of this general type seems capable of correlating satisfactorily at least some of the low temperature thermal properties of rare earth and actinide heavy electron metals[32].

When solved consistently in a Hartree approximation the above quantum Ginzburg Landau model leads to a contribution to the entropy due to thermally exited fluctuations in the magnetisation which may be interpreted as a sum in the Brillouin zone of the entropy of overdamped oscillators[2,32,33] characterised by a relaxation spectrum $\Gamma(\mathbf{q})$ which may be defined via the dynamical wavevector dependent susceptibility determined, for example, by neutron scattering measurements [34-36]. In leading order the entropy shift is linear in temperature and depends only weakly on the bulk average susceptibility in systems described by equations (4) and (5) when c is large, and increases in proportion to the bulk susceptibility when c is small or when $\Gamma(\mathbf{q})$ is nearly independent of q. This basic prediction of the model is consistent with observations in d and f heavy electron metals[2,32].

The model which we have outlined above for the purpose of illustration may be extended to deal with more general (e.g. not necessarily analytic) functionals F[M] of a vector variable $\mathbf{M}(\mathbf{r},\tau)$. The construction of realistic functionals compatible with symmetry, with constraints on **M** and, for example, with inelastic neutron scattering data, is the central problem in this phenomenological approach.

A more complete description of the partition function would also require an explicit account of the contribution of particle density fluctuations, as well as magnetisation fluctuations, and of their mutual coupling. In a multiband model intra-atomic or inter-band excitations lead to local fluctuations in the particle density which can preserve charge neutrality in the unit cell and can therefore be slow enough under suitable conditions to be important in the low temperature thermal properties.

As a guide to developing a realistic Ginzburg Landau free energy functional for f metals we consider the dynamics of a model Hamiltonian of the form

$$\hat{H} = \sum_{k,\sigma} E(k) c_{k\sigma}^{\dagger} c_{k\sigma} + \sum_{j} E_f n_f(R_j) + \sum_{k,j,\sigma} \left\{ V(k) \sqrt{\frac{Q-n_f}{N}} \, c_{k\sigma}^{\dagger} f_{\sigma}(R_j) e^{-ik.R_j} + h.c. \right\}$$

(6)

where $Q = 1$, $N = 2j+1$ is the degeneracy of the atomic f energy levels and the remaining parameters have their conventional meaning. In this approach a strong local constraint is incorporated directly into the Hamiltonian (via the factor $[(Q-n_f)/N]^{1/2}$), rather than into the many particle eigenstates of the Anderson lattice model.

Expanding this model about the mean field values of $\langle c_{k\sigma}^{\dagger} f_{\sigma}(R_j) \rangle$ and of $\langle n_f(R_j) \rangle$, and retaining only terms lowest order in the interactions, we obtain an effective Hamiltonian

$$\hat{H} = \sum_{k,\sigma} E(k) c_{k\sigma}^{\dagger} c_{k\sigma} + \sum_{j} \tilde{E}_f n_f(R_j) + \sum_{k,j,\sigma} \left\{ \tilde{V}(k) c_{k\sigma}^{\dagger} f_{j\sigma} e^{-ik.R_j} + h.c. \right\}$$

(7)

$$+ \sum_{j,\sigma,\sigma'} \left\{ \frac{1}{2} U f_{j\sigma}^{\dagger} f_{j\sigma'}^{\dagger} f_{j\sigma'} f_{j\sigma} - \sum_{k} \left[I(k) c_{k\sigma}^{\dagger} f_{j\sigma}^{\dagger} f_{j\sigma'} f_{j\sigma} e^{-ik.R_j} + h.c. \right] \right\}_{HF}$$

where the subscript j on the f electron operators is a site index. The mean field parameters are

$$\tilde{E}_f = E_f - \frac{1}{Nr} \sum_{k,\sigma} V(k) \langle c_{k\sigma}^{\dagger} f_{\sigma}(R_j = 0) \rangle \,,$$

(8)

$$\tilde{V}(k) = V(k) \sqrt{\frac{(Q-\langle n_f \rangle)}{N}} \,, \qquad Q=1,$$

and the interaction parameters are

$$U = \frac{-1}{2Nr^3} \frac{1}{N} \sum_{k,\sigma} V(k) \langle c^\dagger_{k\sigma} f_\sigma(\mathbf{R}_j = 0) \rangle,$$

(9)

$$I(k) = \frac{1}{2Nr} V(k),$$

where $r = [(Q-\langle n_f \rangle)/N]^{1/2}$ and the subscript HF in equation (7) indicates subtraction of the Hartree Fock contractions of the interactions. The expectation value $\langle c^\dagger_{k\sigma} f_\sigma \rangle$ is negative for a positive $V(k)$, in which case $\tilde{E}_f$ is greater than E_f [37], U is positive so that there is an on-site repulsive interaction, and $I(k)$ is positive so that there is an off-diagonal attractive term in H which plays an important role in properly imposing the constraint.

In the large N expansion, both the exchange contributions and the higher order multi-particle contributions of the interactions are neglected in leading order in $1/N$ [37]. The residual fluctuations about the mean field state are, in the large N limit, high frequency *charge* fluctuations as predicted by a random phase approximation (RPA). [21]

However, in the spirit of our discussion, we regard equation (7) as a model Hamiltonian with weak interactions which we treat in RPA, *without* ignoring the exchange contributions of the interactions. When we do this, we generate *two branches of magnetic fluctuation modes* in addition to the charge fluctuation modes.

We believe that this approach can provide a microscopic basis for a quantum Ginzburg Landau description for f metals and a deeper insight into the mechanism of renormalisation of the quasiparticle spectrum and of the spin dependent quasiparticle interactions which lead to magnetic and superconducting instabilities at low temperatures.

We wish to thank Greg McMullan and Yang Chen for helpful discussions and comments.

References

1. For recent work and references to the literature see *Theoretical and Experimental Aspects of Valence Fluctuations*, eds. L.C. Gupta and S.K. Malik (Plenum, New York, 1987); Proc. of Intern. Symposium on Magnetism of Intermetallic Compounds J. Magn. Magn. Mat. **70** (1987).
2. G.G. Lonzarich, J. Magn. Magn. Mat. **54-57**, 612 (1986); **45**, 43 (1984).
3. L. Taillefer, G.G. Lonzarich, R. Newbury, Z. Fisk and J. Smith, Proc. Conference on Anomalous Rare Earths and Actinides, July 1986, J. Magn. Magn. Mat. **63 & 64**, 372 (1987).

4. P.H.P. Reinders, M. Springford, P.T. Coleridge, R. Boulet and D. Ravot, J. Magn. Magn. Mat. **63 & 64**, 297 (1987); Phys. Rev. Lett. **17**, 433 (1986).

5. K. Andres, J.E. Graebner and H.R. Ott, Phys. Rev. Lett. **35**, 1775 (1975).

6. P. Coleman, J. Magn. Magn. Mat. **63 & 64**, 245 (1987).

7. Z. Fisk, D.W. Hess, C.J. Pethick, D. Pines, J.L. Smith, J.D. Thompson and J.O. Willis, to be published (1988).

8. N. Berk and J.R. Schrieffer, Phys. Rev. Lett. **17**, 433 (1966).

9. S. Doniach and S. Engelsberg, Phys. Rev. Lett. **17**, 750 (1966).

10. W. Brenig, H.J. Mikeska and E. Riedel, Z. Phys. **206**, 439 (1967); E. Riedel, Z. Phys. **210**, 403 (1968).

11. W.F. Brinkman and S. Engelsberg, Phys. Rev. **169**, 417 (1968).

12. M.T. Beal-Monod, S.-K. Ma and D.R. Fredkin, Phys. Rev. Lett. **20**, 929 (1968).

13. K.K. Murata and S. Doniach, Phys. Rev. **29**, 285 (1972).

14. T. Moriya and A. Kawabata, J. Phys. Soc. Japan **34**, 639 (1973); **35**, 669 (1973).

15. C.J. Pethick and G.M. Carneiro, Phys. Rev. **A7**, 304 (1973); **B11**, 1106 (1975); **B16**, 1933 (1977).

16. J.A. Hertz, Phys. Rev. **B14**, 1165 (1976); J.A. Hertz and M.A. Klenin, Physica **91B**, 49 (1977).

17. S.G. Mishra and T.V. Ramakrishnan, Phys. Rev. **B18**, 2308 (1978); Phys. Rev. **B31**, 2825 (1985).

18. A.A. Abrikosov, Physics **1**, 5 (1965).

19. F.D.M. Haldane, Phys. Rev. Lett. **40**, 416 (1978).

20. A. Auerbach and K. Levin, Phys. Rev. Lett. **35**, 3394 (1987).

21. A. Millis and P. Lee, Phys. Rev. **B35**, 3394 (1987).

22. O. Gunnarson and K. Schonhammer, Phys. Rev. Lett. **50**, 604 (1983).

23. N. Read and D.M. Newns, J. Phys. **C16**, 3273 (1983).

24. N.E. Bickers, Rev. Mod. Phys. and references therein, to be published.

25. E. Abrahams, J. Magn. Magn. Mat. **63 & 64**, 234 (1987).

26. B.A. Jones and C.M. Varma, Phys. Rev. **B34**, 6554 (1986).

27. J. Hirsch, Phys. Rev. Lett. **54**, 1317 (1985).

28. J. Hirsch, D. Scalapino and E. Loh, to be published.

29. T. Moriya, *Spin Fluctuations in Itinerant Electron Magnetism*, (New York, Springer, 1985) and references cited therein.

30. Y. Takahashi, *Electron Correlations and Magnetism in Narrow Band Systems*, ed. T. Moriya (New York, Springer, 1981) 91.

31. G.G. Lonzarich and L. Taillefer, J. Phys. **C18**, 4339 (1985).

32. G.J. McMullan, Dissertation, University of Cambridge (1987).

33. G.G. Lonzarich, Summer School on Magnetism in Solids, Kuparovice, Czechoslovakia (1986): unpublished lecture notes).

34. Y. Ishikawa, Y, Noda, Y.I. Uemura, C.F. Majkrzak and G. Shirane, Phys. Rev. B31, 5884 (1985).

35. N.R. Bernhoeft, G.G. Lonzarich, P.W. Mitchell and D.McK. Paul, Phys. Rev. B28, 422 (1983); Physica 136B, 443 (1986).

36. G. Aeppli, A. Goldman, G. Shirane, E. Bucher and M.-Ch. Lux-Steiner, Phys. Rev. Lett. 58, 808 (1987).

37. P. Coleman, Phys. Rev. B35, 5073 (1987).

THE ROLE OF FERMI SURFACE ANISOTROPIES IN HEAVY-FERMION SUPERCONDUCTORS

U. Rauchschwalbe and F. Steglich

Institut f. Festkörperphysik, Technische Hochschule Darmstadt

We present evidence that at the superconducting transition temperature T_c of the Heavy-Fermion-Superconductors $CeCu_2Si_2$ and Ube_{13}, the order parameter is formed only on part of the Fermi-surface. In UBe_{13}, at least part of the Fermi-surface that does not participate in superconductivity at T_c = 0.9K develops an order parameter in the accessible temperature range, i.e. at 0.55K. The effect can be sharply enhanced by doping UBe_{13} with a few at% Th, leading to a second phase transition within the superconducting state. In $CeCu_2Si_2$, those parts of the Fermi surface which do not participate in superconductivity down to the lowest temperatures carry a small effective mass. The influence of impurities on the Fermi-surface anisotropy is discussed.

In 1985, Ott et al. reported the existence of a second phase transition that shows up below the superconducting one in UBe_{13} samples doped with a few percent of Th [1]. Here, we will briefly review the explanation given in Ref. 2 for this exciting phenomenon, and then discuss its implication for the understanding of the superconducting (s.c.) state of UBe_{13} and also of the Ce-based Heavy-Fermion Superconductor $CeCu_2Si_2$. In Fig. 1b we reproduce

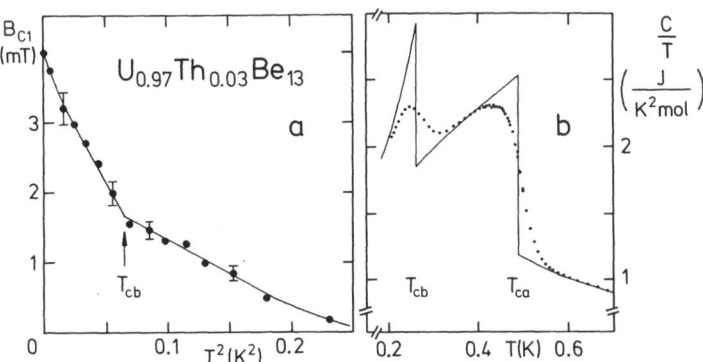

Fig. 1. (a) Lower critical field of $U_{0.97}Th_{0.03}Be_{13}$ in a plot B_{c1} vs. T^2. Solid line is a guide to the eye.
(b) Specific heat of the same sample.[7] Solid line is a schematic replacement of the data by two sharp transitions.

the specific heat results on a $U_{0.97}Th_{0.03}Be_{13}$ sample[2], which exhibits two phase transitions, the superconducting one at $T_{ca} = 0.48K$ and the second one at $T_{cb} = 0.26K$. Fig. 1a displays results on the lower critical field $B_{c1}(T)$. Most remarkable is the abrupt increase of $B_{c1}(T)$ below T_{cb}, which corresponds to an increase of the superfluid density. The results in Fig. 1 allow to discard the idea[3] that at T_{cb} one (non-conventional) s.c. order parameter is **replaced** by another one, as it happens in superfluid ^{3}He. This could be the case if:

    i)   the transition at T_{cb} were of first order. If one for the moment assumes this to be, one can calculate the latent heat associated with such a transition, using **only** the measured $B_{c1}(T)$ and the Ginzburg-Landau-parameter $\kappa \simeq 30$ [2]. This yields $\Delta Q \simeq 50$ mJ/mol or, at $T_{cb} \simeq 0.25K$, an entropy change of $\Delta S \simeq 0.2J/Kmol$, which corresponds to the total measured entropy between 0.2K and 0.3K, clearly ruling out i) and leaving:

    ii)   a second order transition from superconductor A to superconductor B can only happen if $B_{c1}(T)$ vanishes at T_{cb}, which is obviously not the case.

Hence, another mechanism must cause the transition at T_{cb}: Upon cooling, a s.c. order parameter Δ_a is formed at T_{ca}. At T_{cb}, a second order parameter Δ_b **adds** to Δ_a. At $T < T_{cb}$, Δ_a and Δ_b coexist. Such a situation is in principle known from the work by Shiffman et al.[4] on Pb-Sn eutectic alloys. While, however, in these eutectics the two order parameters coexist in different **spatial** regions of the sample, in $U_{0.97}Th_{0.03}Be_{13}$ this coexistence appears to be in different regions of k-space, i.e. on different parts of the Fermi-surface.

These different regions are coupled by scattering of Cooper pairs from one part of the Fermi-surface into the other, causing the observed broadening of the C(T)-transitions. This is therefore an **intrinsic** effect and cannot be related to sample inhomogeneities.

In the following we will argue that two transitions can also be observed in pure UBe_{13}, but that the order in which they appear has reversed: Pro-

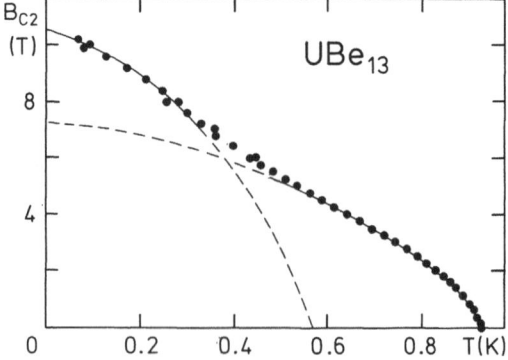

Fig. 2.   Upper critical field $B_{c2}(T)$ of UBe_{13}. The solid lines are guides to the eye, the dashed lines are extrapolations, assuming that $B_{c2}(T)$ is determined by two superconducting order parameters (see text and ref. 8).

nounced maxima in the ultrasound attenuation have been observed in $(\underline{U},Th)Be_{13}$ below T_{cb}[5] and in pure UBe_{13}[6] below $T_c =: T_{cb}$. The specific heat of s.c. UBe_{13} is enhanced compared to simple model calculations[7]. This excess specific heat can be regarded as a broadened phase transition around $T_{ca} \simeq 0.55K$[8]. The occurrence of two order parameters is also apt to explain the strange upper critical field of UBe_{13} (see Fig. 2): The order parameter Δ_b formed at $T_{cb} \simeq 0.9K$ has a smaller critical field than Δ_a ($T_{ca} \simeq 0.55K$), leading to the observed upswing in $B_{c2}(T)$. The above point of view also explains the dependence of T_{cb} on Th-concentration x.[8] For $x \leq 0.5\%$, $T_{ca} < T_{cb}$, and scattering of Cooper pairs belonging to Δ_b happens into normal regions of the Fermi surface. This is an efficient pair-breaking mechanism which explains the rapid decrease of $T_{cb}(x)$, until at $x > 0.5\%$, $T_{cb} < T_{ca}$. Now, the scattering is always into s.c. regions and $T_{cb}(x)$ levels off. Note that this requires Δ_a and Δ_b to have a similar symmetry.

The band structure calculations for $CeCu_2Si_2$ and $CeAl_3$ by Sticht et al.[9] show that the formation of the Heavy-Fermion state goes along with large anisotropies of the effective mass over the Fermi-surface. Therefore, the picture sketched above is not as surprising as it may look at first glance, if one accepts that regions on the Fermi-surface with different effective masses have different coupling constants. One might even expect that the other Heavy-Fermion superconductors display similar phenomena. To elucidate this point, we have performed a systematic investigation of the upper critical field $B_{c2}(T)$ of a number of $CeCu_2Si_2$ samples with different

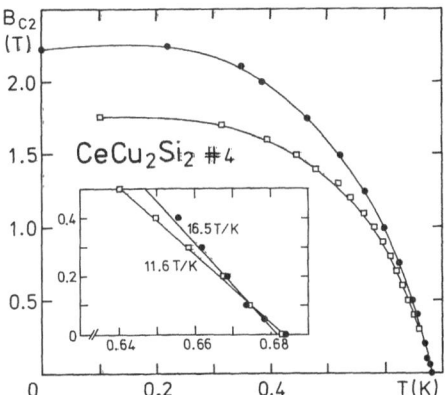

Fig. 3. $B_{c2}(T)$ of a $CeCu_2Si_2$ sample[7] in two different annealing stages; ( ): unannealed, $\rho_0 = 60\mu\Omega cm$, ( ): annealed 4d, 1000°C, $\rho_0 = 12\mu\Omega cm$.

residual resistivities ρ_0. The initial slope of $B_{c2}(T)$ is in the clean limit given by:

$$B'_{c2}(\rho_0=0) = 1.38\cdot10^{35}(Tm^2K^2/J^2)\cdot\gamma^2 T_c/S_s^2, \qquad (1)$$

where γ is the coefficient of the electronic specific heat and S_s is the area of the Fermi-surface the Cooper pairs are formed on. Taking $B'_{c2}(\rho_0=0)$ $\simeq$ 10T/K from a B'_{c2} vs. ρ_0 plot, we calculate $S_s \simeq 13 \cdot 10^{20} m^{-2}$ which is only half of the total Fermi-surface $28 \cdot 10^{20} m^{-2}$ determined at elevated temperatures from the maximum value of the electrical resistivity.[7] Hence, also in $CeCu_2Si_2$ only half of the Fermi surface participates in the superconductivity at T_c $\simeq$ 0.6K. In contrast to $(\underline{U},Th)Be_{13}$, however, the rest of the fermi-surface does not seem to form Cooper pairs in the investigated T-range (down to about 30 mK). This may be related to the fact that the electrons on these parts carry only a small effective mass as is evidenced by a small residual $(T \rightarrow 0)$ γ-term of the order of 30mJ/K^2mol in the best samples,[7] while simultaneously a large linear term in the thermal conductivity points to light electrons with a large Fermi velocity.[10] Possibly, a minimum effective mass is required to form Cooper pairs in these compounds; a conclusion that has been realized before.[2] It should be noted that in one exceptionally pure $CeCu_2Si_2$ sample, indications for the presence of a second phase transition at about $0.5 \cdot T_c$ have been observed.[7]

This leads us now to extend our considerations on $CeCu_2Si_2$ away from the clean limit. It has been demonstrated that B'_{c2} in this compound increases with increasing ρ_0 (see e.g. the data in Fig. 3). Such a dependence has to be expected from the full expression for B'_{c2} for an isotropic conventional superconductor:

$$B'_{c2} = 1.38 \cdot 10_{35}(Tm^2K^2/J^2)\gamma^2 T_c/S^2_s + 4490(Tm^2K/\Omega J)\gamma\rho_0 \qquad (2)$$

However, the **average** increase of B'_{c2} with ρ_0 as determined on about 20 samples with different ρ_0, $\partial B'_{c2}/\partial\rho_0 \simeq 0.15$ T/$K\mu\Omega$cm, is only 20% of what one calculates using a typical $\gamma \simeq 0.7$J/K^2mol. Two possible reasons for this shall be discussed now:

i)   In analogy to $(\underline{U},Th)Be_{13}$, scattering can be expected to act as a pair-breaking mechanism for the highly anisotropic s.c. state of $CeCu_2Si_2$ thus giving a negative contribution to B'_{c2} and partly cancelling the positive one proportional to ρ_0. then, one would expect T_c to be reduced as well, a tendency that is not very pronounced with our samples, but has been observed on radiation damaged samples.[11]

ii) The scattering may lead, via induction, to an increase of S_s, which would also reduce B'_{c2}.

To summarize, we have shown that in the Heavy-Fermion Superconductors $(\underline{U},Th)Be_{13}$ and $CeCu_2Si_2$, the s.c. order parameter is at T_c only formed on part of the Fermi surface, giving rise to an anisotropic superconducting state. We wish to emphasize that our considerations are independent of the type of Cooper pairing present. They can, therefore, not help to decide whether the order parameters in UBe_{13} and $CeCu_2Si_2$ are conventional or nonconventional ones.

We gratefully acknowledge close collaboration with P. Fulde, K. Maki, C.D. Bredl and G. Sparn. This work was supported by the Sonderforschungsbereich 252 Frankfurt-Darmstadt-Mainz-Stuttgart.

REFERENCES

1.   H.R. Ott, H. Rudigier, Z. Fisk, and J.L. Smith. <u>Phys</u>. <u>Rev</u>. B31:1651 (1986).
2.   U. Rauchschwalbe, F. Steglich, G.R. Stewart, A.L. Giorgi, P. Fulde, and K. Maki. <u>Europhys</u>. <u>Letters</u> 3:751 (1987).

3. R. Joynt, T.M. Rice, and K. Ueda. _Phys. Rev. Lett._ 56:1412 (1986).
4. C.A. Shiffman, J.F. Cochran, M. Garber, and G.W. Pearsall. _Rev. Mod. Phys._ 36:127 (1964).
5. B. Batlogg, D.J. Bishop, B. Golding, C.M. Varma, Z. Fisk, J.L. Smith, and H.R. Ott. _Phys. Rev. Lett._ 55:1319 (1985).
6. B. Golding, D.J. Bishop, B. Batlogg, W.H. Haemmerle, Z. Fisk, J.L. Smith, and H.R. Ott. _Phys. Rev. Lett._ 55:2479 (1985).
7. U. Rauchschwalbe, U. Ahlheim, C.D. Bredl, H.M. Mayer, and F. Steglich. _J. Magn. Magn. Mat._ 63&64:447 (1987).
8. U. Rauchschwalbe, C.D. Bredl, F. Steglich, K. Maki, and P. Fulde. _Europhys. Lett._ 3:757 (1987).
9. J. Sticht, N. d'Ambrumenil, and J. Kübler. _J. Magn. Magn. Mat._ 63&64:254 (1987).
10. F. Steglich, C.D. Bredl, W. Lieke, U. Rauchschwalbe, and G. Sparn. _Physica_ 126B:1982 (1984).
11. G. Holter and H. Adrian. _Solid State Comm._ 58:45 (1986).

ORDERING PHENOMENA IN HEAVY-FERMION COMPOUNDS

J.J.M. Franse

Natuurkundig Laboratorium der Universiteit van Amsterdam

Valckenierstraat 65, 1018 XE Amsterdam, The Netherlands

The heavy-fermion systems are characterised by anomalies in thermodynamic and transport properties at temperatures typically below about 10 K. These anomalies indicate an effective mass of the electrons at the Fermi level, which is two orders of magnitude larger than the free-electron mass. Although the large effective mass is intriguing in itself, even more astonishing is the occurrence of superconductivity in the heavy-fermion compounds as well as the extremely large specific-heat anomalies at the superconducting transition temperature, indicating that heavy-mass electrons participate in the superconductivity. These features have given rise to suggestions of an exotic type of pair bonding between the electrons. In the familiar models of superconductivity the necessary attractive interaction between two electrons is provided by electron-phonon processes. The most favourable situation occurs for a symmetrical orbital function of the electron pair with opposite spins (s-wave or singlet pairing). Moreover, the Debye temperature that represents the range of phonon energies should be much smaller than the Fermi temperature of the electrons. In normal metals this condition is well fulfilled since the Fermi temperature is of the order of 10^4 to 10^5 K whereas the Debye temperature is a few hundred kelvin. As the Fermi temperature is inversely proportional to the electronic mass its value in the heavy-fermion compounds is reduced to a value comparable to the Debye temperature. From this point of view, pair bonding in the superconducting state by attractive electron-electron interactions not intermediated by electron-phonon processes is a real possibility. At present, an attractive interaction intermediated by so-called spin fluctuations is considered to be a good candidate. The bonding of an electron pair is now connected with the spin polarization of the electron gas due to magnetic interactions. These interactions dress the electron with a cloud of polarized spins. This dressing of the electron is the origin of its high effective mass. One conjecture is that the most favourable pairbonding occurs in this case for electrons with parallel spins and hence with an antisymmetric orbital function (p-wave or triplet pairing). The energy range of the spin fluctuations is now represented by the spin-fluctuation temperature, T_{sf}, which appears to be of the order of 10 K, thus small compared to the effective Fermi temperature. Since in this picture magnetic interactions play an important role, a competition between superconductivity and magnetic order must be expected.

A somewhat arbitrary limit for characterising a compound as a heavy-fermion system is given by a value of 400 mJ/K^2mol for the electronic coefficient γ of the linear term in the specific heat/temperature relation.

(In normal metals the corresponding figure is between 1 and 10). A number of compounds are presently known to meet this criterion: $CeAl_3$, $CeCu_2Si_2$, $CeCu_6$, UBe_{13}, UPt_3, U_2Zn_{17} and UCd_{11}. Three of them become superconducting below 1 K ($CeCu_2Si_2$, UBe_{13} and UPt_3), two exhibit antiferromagnetic order below 10 K (U_2Zn_{17} and UCd_{11}) and the remaining two show no order at all down to the millikelvin temperature range ($CeAl_3$ and $CeCu_6$). In general, the heavy-fermion behaviour in these cerium and uranium intermetallics is approached in terms of spin-fluctuation, valence-fluctuation or Kondo-lattice models. In the present contribution the attention is focussed on the first model, already introduced above.

Ordering phenomena in the heavy-fermion systems are sometimes extremely sensitive to deviations from stoichiometry, to chemical purity, to the atomic order or to the state of stress. In $CeCu_2Si_2$ compounds which are copper-rich, superconductivity is enhanced, whereas in UPt_3, the superconducting temperature, T_s, is strongly dependent on annealing procedures. Besides that, superconductivity in UPt_3 is easily destroyed by powdering the material or by substitution of only 0.5 at% Pt by iso-electronic Pd. On further increasing the palladium content in UPt_3 the heavy-fermion regime subsists up to 10 at%, the alloys first showing spin-fluctuation effects and then, at between 1 and 2 at% Pd, an onset to antiferromagnetic order with maximal values for the Néel temperature of about 6 K around 5 at% Pd.

SPIN FLUCTUATIONS

Increasing pressure depresses the low-temperature anomalies that are characteristic for the heavy-fermion behaviour. This depression has been demonstrated for UPt_3 in resistivity and specific heat measurements. Both the coefficients A of the T^2-term in the resistivity and the coefficient γ of the linear term in the specific heat/temperature relation decrease with pressure. The parameter A/γ^2 turns out to be almost pressure independent and its value is found to be close to 1.0×10^{-5} $\mu\Omega cm$ (mole K/mJ)^2, a common value for nearly all heavy-fermion compounds. This universal result enhances the challenge to bring these systems into the same theoretical framework. Working within the concept of spin fluctuations, the parameter A is found to be inversely proportional to the square of the spin-fluctuation temperature. According to the experimental observations for UPt_3, the coefficient γ varies with pressure as $1/T_{sf}$. Pressure effects on the susceptibility of UPt_3 can also be expressed in terms of a pressure dependent spin-fluctuation temperature. Numerical results for the relative dependence of T_{sf} as determined from resistivity and susceptibility measurements under pressure, coincide within experimental error. This feature, however, should not be considered as a proof that spin fluctuations are on the basis of the heavy-fermion behaviour of this particular compound.

SUPERCONDUCTIVITY

Both negative and positive pressure effects have been reported for the superconducting transition temperature in heavy-fermion compounds. For example, in the uranium-based heavy-fermion superconductors, UBe_{13} and UPt_3, the pressure dependence of T_s is moderately negative and comparable to what is usually observed for a conventional type of superconductor. The most interesting feature, however, is not the sign or absolute value so much as the relation between pressure effects in the normal and superconducting states. Studies of the pressure dependence of specific heat and resistivity can provide us with the required information on such normal-state properties like the heavy electronic mass and the Kondo or spin-fluctuation temperatures, but interpretation of the data depends strongly on appropriate models which are not always available; the temperature dependences that are observed in these materials are often complex.

Since spin fluctuations are known to suppress superconductivity in normal metals, the coexistence of both phenomena in UPt_3 was certainly not expected. A start to interpreting the superconductivity and its pressure dependence for this compound has been made by Pethick et al. Working from a Fermi liquid model, the normal and superconducting state parameters are expressed in terms of the same Landau parameters. Values for these parameters derived in the normal state indicate a p-wave-type of superconductivity and predict a satisfactory correspondence with the experimentally observed pressure dependence of the superconducting transition temperature. This promising approach is applicable to UPt_3 because an adequate fit of the specific heat data is given by Fermi liquid theory that includes a $T^3 \ln T/[K]$-term. A similar description of the other heavy-fermion superconductors is lacking at present.

ANTIFERROMAGNETIC ORDERING

Antiferromagnetic order in the heavy-fermion systems has been observed for U_2Zn_{17} and UCd_{11} with values for the Néel temperature, T_N, of 9.7 and 5.0 K, respectively. Substitution of two percent Zn by Cu depresses the Néel temperature of U_2Zn_{17} from 9.7 K down to below 1.5 K, whereas substitutions on the uranium sublattice have a much weaker effect. On the other hand, impurity effects on the specific heat above the ordering temperature are small in U_2Zn_{17} as in UPt_3. Although in UP_3 no long-range magnetic order has been found, high-field magnetization measurements show a metamagnetic-type of transition in this material around 20 T in the liquid-helium temperature region, reminiscent of some type of antiferromagnetism. Upon substituting a few percent of Pt by Pd or U by Th, this latent antiferromagnetism becomes apparent below an ordering temperature of about 6 K. Experiments indicate that antiferromagnetism in these alloys is associated with Fermi surface instabilities (spin-density waves). High-pressure experiments reveal a strong depression of T_N in $U(Pt,Pd)_3$ compounds in contrast to U_2Zn_{17} and UCd_{11} where small and positive pressure effects have been reported. In these latter cases, the positive pressure effects are claimed to result from competing Kondo and indirect exchange interactions.

BASIC MECHANISMS

It seems tempting to ascribe a large substitution effects on ordering phenomena in the heavy-fermion compounds either to impurity scattering, to lattice spacing effects or to critical changes in 5f-electron localization or hybridization between the uranium and surrounding ligand states. Taking these in turn, in our experience, impurity scattering is not the dominant parameter in suppressing superconductivity in the (U, Th)$(Pt,Pd)_3$ system. Unannealed UPt_3 samples with $\rho(300 \text{ K})/\rho_0$-values of 30 are superconducting below 0.3 K, whereas in a 1 at% Pd sample with the same value for this resistivity ratio, no superconductivity has been observed down to 40 mK. Because the molar volume of UPt_3 increases with increasing thorium but decreases with increasing palladium content, the atomic volume can not be the decisive parameter for the transition from the superconducting to the magnetically ordered state in the (U, Th)$(Pt,Pd)_3$ system either. Further parameters to be considered are the lattice constants. The c/a ratio for the hexagonal compound UPt_3 increases on substituting with thorium or palladium. This suggests that the depression of superconductivity and promotion of antiferromagnetic order goes with decreasing c/a values, and since this ratio increases with pressure (the compressibility is slightly anisotropic in UPt_3), the conclusion is consistent with the results of the high-pressure studies on the antiferromagnetic ordering temperature. However, it does not explain the depression of superconductivity by both high-pressure and increasing thorium or palladium content. Substitutions on the uranium sub-lattice (thorium) as well as on the platinum sublattice (palladium) most probably lead to increasing localization of the 5f electrons due to increased U-U distances or to a reduced f-d hybridization, respectively. This increased localization is undoubtedly in favour of a magnetically ordered state and compatible with the negative pressure effect on T_N. The negative pressure effect on T_s, however, remains again unexplained. We have to admit that our present understanding of the mechanisms that drive the (U, Th)(Pt,Pd)$_3$ compounds from the superconducting state into the magnetically ordered (spin-density-wave) state is still poor.

CONCLUSION

To conclude this examination of ordering phenomena in the heavy-fermion systems a tentative phase diagram of ordering phenomena in the pseudo-binary compound U(Pt,Pd)$_3$ is discussed. Up to a palladium concentration of 10 percent, these compounds remain in the heavy-fermion regime with values of the electronic-mass enhancement of 180 or more. Within this range, the low-temperature anomalies in the resistivity change from spin-fluctuation-type to Kondo-type. The ordering phenomena change from superconductivity to antiferromagnetic order with a narrow concentration range around 1 at% Pd where no ordering has been observed. The spin fluctuation temperature increases with the application of pressure, whereas the superconducting and antiferromagnetic ordering temperatures are depressed, leaving open the possibility that the same interactions are responsible for supercon-ductivity and antiferromagnetism. Within this series of pseudo-binary com-pounds we thus meet all characteristic phenomena of heavy-fermion systems that have been observed: spin-fluctuation effects, Kondo behaviour, coherence effects, superconductivity, antiferromagnetism, spin-density waves and lack of an ordered ground state. Uniaxial stress, hydrostatic pressures and internal (chemical) pressures provide the experimentalists with a power-ful arsenal for unravelling the intriguing low-temperature properties of the heavy-fermion systems.

REFERENCES

A. Ponchet, J.M. Mignot, A. de Visser, J.J.M. Franse, and A. Menovsky, J. Magn. Magn. Mat. 54-47:399 (1986).
G.E. Brodale, R.A. Fisher, N.E. Philips, G.R. Stewart, and A.L. Giorgi, Phys. Rev. Lett. 57:234 (1986).
C.J. Pethick, D. Pines, K.F. Quader, K.S. Bedell, and G.E. Brown, Phys. Rev. Lett. 57:1955 (1986).
A. de Visser, Thesis, University of Amsterdam (1986).

GAINING CONTROL OVER RARE EARTH VALENCE FLUCTUATIONS

Dieter Wohlleben[*]

Center of Materials Sciences

Los Alamos National Laboratory, Los Alamos, NM 87545, USA

SOME WORKSHOP-STYLE REMARKS ON THE PROBLEM OF NARROW BAND MATERIALS

The problem addressed in this workshop is in my view a very old one. It came up for the first time in the early thirties with the discovery of the resistivity minimum in dilute alloys of simple metals with magnetic transition metal impurities. If this view is correct, then the existence of this workshop in 1987 proves that the problem turned out to be a very difficult one because it is apparently not yet solved to anybody's satisfaction in spite of the vast amount of experimental and theoretical work done so far.

In the course of time many names have been attached to the field: Kondo effect, spin fluctuations, interconfigurational fluctuations, valence fluctuations, mixed valence, narrow band phenomenon, f instability, Anderson lattice, heavy fermion phenomenon, strongly correlated systems, dynamic alloy etc. Curiously, all these names have been applied to the very same physical system at some time or another (e.g. to $CeAl_3$). If nothing else, all these names then reflect a sequence of hope and despair and finally a great amount of confusion; it is in fact sometimes very difficult, even for an old hand like me, to keep the basic problem in view through this fog of semantics.

The names derive either from a particular theoretical model, which addresses a limited aspect of this problem (Kondo effect, spin fluctuations, Anderson lattice) or from a particularly striking experimental observation like the large electronic specific heat coefficient (narrow band and heavy fermion phenomenon) or from a desire to keep the entire problem in view, at the cost of being unable to lean on a crisp but inherently limited i.e. unrealistic theoretical model (valence fluctuations).

When then is the basic problem? There is no doubt in anybody's mind, that the phenomena of interest arise from an instability of the partially filled d or f shell of certain atoms when they are put into a metallic host. The theoretical models which dominate the scene work with two local d or f states on one hand (they have now actually gone beyond a local s electron!) and a structureless sea of free conduction electrons on the other.

---

[*] on leave of absence from II. Physikalisches Institut der Universitaet zu Koeln, 5000 Koeln 41, FRG.

In my view this procedure ignores at least half of the essential physics; the other half is kept alive in the term valence fluctuation. Basically, what the prevalent models ignore, is that in all these systems we are actually dealing with entire _atoms_ as the source of the anomalies, not just with their f shells. In other words, there _is_ important structure in the sea of conduction electrons.

If one assumes charge neutrality of the RE unit cell in the metal (a very reasonable first approximation), the creation of a hole in the f shell implies the creation of a new electron on the valence electron shell of the _same_ atom after a time of order h/ε_F. This time is very short compared to the lifetime of the f hole in any narrow band material and therefore the existence of the new valence electron has drastic effects for the equilibrium properties; it should not be left out of the Hamiltonian, as it is in the Anderson model. The new valence electron is also a conduction electron, but the conduction electron density is drastically different before and after the creation of the f hole on this particular RE atom. Early attempts to include this electron were made by Falicov and Kimball[1] and by Kaplan and Mahanti[2], but a vigorous follow through along this line has unfortunately not materialized.

The most important consequences of the existence of this new valence electron are:

(a) A change of the binding energy of the atom to its environment. I do not understand why this energy must be buried in some effective ff Coulomb repulsion. The total energy of the transition is, at zero mixing matrix element, the sum of a well known f-d-transfer energy (about 2.6 eV in the di- to trivalent transition of RE atoms) and of the difference of the atomic binding energies to the environment which depend on the particular solid and which can be measured by high energy spectroscopies (one of their two foremost tasks, in my view) or can be calculated in Miedema's scheme[3]. Why abandon control over this most important energy parameter for some models sake?

(b) A drastic change of the charge distribution in the RE atom. This means, that a slow f spin fluctuation is necessarily accompanied by an intraatomic charge fluctuation which is just as slow.

(c) The scattering cross-section for conduction electrons of an electron in the valence electron shell of a RE atom is nearly two orders of magnitude larger than that of an electron in the f shell. The resistivity anomalies are therefore dominated by the forgotten extra valence electron (4).

(d) A change of the radius of the RE atom, which will induce displacements of the atoms (nuclei) of the nearest neighbor shell and beyond, if the lifetime τ_h of the f hole is longer than $h/k_B\Theta_D$, where Θ_D is the Debye temperature. In the most interesting narrow band materials we do have $\tau_h > h/k_B\Theta_D$.

(e) The charge distributions of most states of the open f and d shell are aspherical, and there is usually strong and always at least residual spin orbit coupling in all these shells. Therefore, if this aspherical charge distribution is involved in a slow charge fluctuation ($\tau_H > h/k_B\Theta_D$), the f or d magnetic moment may be demagnetized by the electron-phonon interaction, rather than by interaction with the conduction electrons. I fail to see why this important possibility has been ignored for so long, until recently.[5] This example shows that one may actually solve the wrong problem when trying to demagnetize the local f or d magnetic moment via the Anderson or Kondo lattice models in narrow band materials. In fact, Noziere argues that this cannot be done at all[6].

One may state categorically, that the rich phenomenology of the systems in question as function of temperature, pressure, and composition will never come under quantitative control, if one keeps ignoring what is going on in the RE valence electron shell and in the phononic system of narrow band materials. It is hard for me to understand why progress in addressing the _entire_ problem has been so slow. We are constantly promised by the theoreticians, that once they have solved their limited problem of the Anderson model, they will include the rest of the physics into the Hamiltonian. But the extra valence electron cannot be treated as a perturbation on the Anderson Hamiltonian. Therefore, what will be the use of all those beautiful solutions of the Anderson model in the ground state vis a vis reality? (The success of the Anderson model in explaining the high energy excitations in photoemission and in BIS[7] seems to say merely, that one can forget the extra valence electron in these fast measurements).

One wonders what would happen if somebody would start again from scratch, for instance with an atomic configurational model similar to that of Hirst[8], but including the valence electron configuration together with the f shell configuration. Should the theoretical chemists not get into the game here, finally?

THE GIBBS FREE ENERGY OF VALENCE FLUCTUATORS

Some of us experimentalists who have been sitting on a vast amount of ground state data for a very long time and are tired of churning out more of the same, are beginning to look for ways out of the dilemma of being unable to process these data sensibly on the basis of existing theoretical results. One way out is to anticipate the results of the more comprehensive theory, necessary to really solve the problem by constructing phenomenologically an expression for the free energy of these systems as function of temperature and pressure with states whose energies and degeneracies are measurable. In the following I shall describe the main results of our work in Cologne in this direction done over the last few years. By trial and error fits to the data we have constructed a free energy expression which describes by now very well the susceptibility, the volume, the neutron quasi-electric linewidth, the L_{III} valence, and the resistivity as function of temperature and pressure for elemental Ce and for RE Cu_2Si_2 with RE = Ce, Eu and Yb. Basically, what we were able to achieve, is to calculate the properties of these narrow band materials from the measurable properties of the two metals with adjacent integral valence.

We use an expression for the Gibbs free energy of the unit cell of the system with six terms[9,10,11,12]

$$G = F_C + F_S + F_E + F_M + F_P + pV$$

All these terms depend on volume V and/or on the valence, i.e. the ratio $v/(1-v)$ in which two adjacent integral valent atomic configurations are present. We minimize G with respect to V and v at fixed temperature T and pressume p.

F_C, the chemical term, contains the basic f-d transfer energy and the basic difference in chemical binding energies of the two valence states in their respective compounds at some fixed T and at p = 0. It is a linear function of v.

F_S, the solution term, takes account of the energy of solution of a very dilute atom with minority valence in a matrix with majority valence. It is important at nearly integral valence; without it, all γ type Ce systems would e.g. be strictly trivalent, i.e. would show no f instability whatsoever.

F_E, the elastic term (also called the alloy segregational term), takes into account the alloy strain energy which follows from the difference of the charge densities on the valence electron shell in the two valence states. This term is zero at integral valence and large and positive at fractional valence. At fixed fractional valence, it is large when the f hole lifetime is short ($\tau_h < h/k_B \Theta_D$, up to 0.4 eV in α Ce), i.e. when the nuclei in the environment cannot follow the charge fluctuation, but becomes very small at long f hole lifetime (in γ Ce and in very narrow band or heavy fermion situations at small T), when the neighborhood can relax elastically. This term and its drastically different behaviour at long and short τ_h is crucial in the $\gamma - \alpha$ transition of all Ce systems which we have tested so far.

F_M is the mixing term. At high temperature and long hole lifetime (weak mixing) it corresponds exactly to the entropic term $- TS$ of the free energy of a mixture of two species of noninteracting RE atoms with their respective crystal field and spin orbit splittings and degeneracies. At low temperature this term also includes the energy gain by quantum mixing between these states, which we calculate by a scheme described repeatedly elsewhere[9,10,11,13]. We have found that the difference between heavy and bantam fermion systems (e.g. between $CeCu_6$ and $CeBe_{13}$) does not come from a difference in fractional valence, as we all expected naively in the past, but from a difference in mixing: At low temperature (thermal energy small compared to crystal field splitting) in bantam fermion systems the groundstates of the two valence states mix as strongly as at elevated temperature, while in heavy fermion systems the mixing almost ceases when k_BT drops far below the CF splitting. This means, that in general the mixing matrix element depends on the pair of configurational substates which it connects, that there are weakly and strongly mixing pairs, and that the special feature of heavy fermion systems (and probably also of RE impurities with small "Kondo" temperatures) is that in these the groundstates are very weakly mixing. We also mention that all matrix elements depend on volume[14,9,15].

F_p is the phonon term. The frequency of the local Einstein oscillator represented by the RE atom depends on its valence state (on the local valence electron charge density and on the binding energy to the environment). This term causes e.g. the different thermal expansion coefficients of RE compounds with different valence. Without this term it turns out to be impossible to track by calculation the temperature dependence of the valence as observed by L_{III} X-ray absorption or by temperature dependent X-ray diffraction.

Obviously the number of energy terms necessary to describe the temperature and pressure dependence is large, much larger than anticipated in the prevalent theoretical models. However, everyone of these terms makes physical sense and nearly all energies necessary for the calculations can be obtained by independent measurements on integral valent reference compounds or by inelastic neutron scattering on the system in question. There is no doubt that the problem is very complicated and that a lot of data are necessary on each individual system to gain quantitative control. There is, however, also no doubt in my mind that the problem can be and has been solved in practice without waiting for higher theory to come down to reality.

ACKNOWLEDGEMENTS

The author thanks J.L. Smith and S. Hecker for supporting his stay at Los Alamos. B. Wittershagen did most of the tedious thermodynamical calculations; without his untiring effort, the results described above would never have materialized. The work in Cologne was supported by SFB 125, Deutsche Forschungsgemeinschaft. Finally I thank my wife, Nicolette Wohlleben, for her nice typing job.

REFERENCES

1. L.M. Falicov and J.C. Kimball, Phys. Rev. Lett. 22:997 (1969).
2. T.A. Kaplan and S.D. Mahanti, Phys. Rev. Lett. 51A:265 (1975).
3. A.R. Miedema, P.F. de Châtel and F.R. de Boer, Physica 100B:1 (1980).
4. D. Wohlleben and B. Wittershagen, Adv. in Physics 34:403 (1985).
5. H. Capellman and K.U. Neumann, preprint, April 1987.
6. P. Nozieres, Ann. de Phys. 10:19 (1985).
7. J.W. Allen, this volume.
8. L.L. Hirst, Phys. Kond. Mat. 11:255 (1970).
9. J. Roehler, D. Wohlleben, J.P. Kappler, and G. Krill, Phys. Lett. 103A:20 (1984).
10. D. Wohlleben, pq. 171 ff in "Moment Formation in Solids" W.L. Buyers, editor, Nato ASI Series B 177, Plenum New York, London (1984).
11. D. Wohlleben, pg. 85 ff in "Physics and Chemistry of Electrons and Ions in Condensed Matter", J.V. Acrivos, A. Joffe, and N.F. Mott, eds. D. Reidel Publ. Co., Dordrecht (1984).
12. B. Wittershagen and D. Wohlleben, to be published.
13. D. Wohlleben and B. Wittershagen, J. Mag. Mag. Mat. 52:32 (1985).
14. J.W. Allen and R. Martin, Phys. Rev. Lett. 49:1106 (1982).
15. D. Wohlleben and J. Roehler, J. Appl. Phys. 55:1904 (1984).

NARROW LINEWIDTH PHENOMENA OF DILUTE d AND f ATOMS IN METALS

D. Riegel[*]

II. Physikalisches Institut, Universitat Köln

5000 Köln 41

GENERAL TRENDS, PROBLEMS

The magnetic behaviour and electronic structure of a dilute transition metal ion in a host metal is the result of a competition between the tendency towards localization of d or f shells, and charge and spin delocalization due to hybridization[1]. The study of dilute systems is necessary to improve our knowledge about the basic interactions between the d or f orbitals with conduction electrons and/or ligands. It is also important for an understanding of narrow band phenomena in concentrated systems.

For several years I have been interested in the magnetic behaviour of 4f, 3d and 4d atoms under extreme changes of the chemical environment. Some of the essential aspects of our work have been the production and investigation of extremely delocalized 4f shells and extremely localized 3d- and 4d shells in metallic systems.

Can one find nearly nonmagnetic Pr systems? Can one produce d ion-host combinations, in which the 3d and 4d shells behave nearly localized? How large are the spin linewidths and crystal fields? What are the ground states? What are the conditions and mechanisms which lead to 3d and 4d shell localization in metals? What are the similarities and differences of magnetic 3d and 4d systems compared to 4f systems? Why is the 3d magnetism usually predominated by spin only magnetism in contrast to the ionic-like behaviour in f systems?

METHOD

Parts of these questions address extreme cases of local magnetism[2] and can be studied in non-alloying systems only. Such systems can be produced by heavy ion reactions and recoil implantation. The local susceptibility and spin dynamics of the implanted 4f or d ions can be measured by the time differential perturbed angular γ-ray distribution (TDPAD) method. In the context of this paper I would like to note the following features of the method: (a) One can observe the magnetic single-ion-behaviour since the

---

* Permanent address: Fachbereich Physik, Freie Universitat Berlin, Institut fur Atom- und Festkörperphysik, 1000 Berlin 33.

concentration of the f or d ions produced is less than 1 ppm. (b) The implantation technique permits the study of local moment formation in many simple hosts, especially in non-alloying systems. (c) For selected cases, the spin dynamics of the isolated ions can be observed in s, sp, d, and f metal hosts, so that one can obtain information about the hybridization as a function of the character of the host conduction electrons.

STRONGLY HYBRIDIZED 4f STATES IN METALS

Magnetic instabilities of single Ce, Pr, and even of Nd and Pm ions have been produced by implantation of these ions in small-volume d hosts like Ta, W, Re, Os, Pt.[3,4] It is relatively easy to produce nonmagnetic Ce systems[3], and it is possible to produce nearly nonmagnetic Pr systems[4]. The 4f-host d hybridization seems to be strongest in d hosts in the middle of the d metal series.

Clear systematic trends in the susceptibility and 4f spin dynamics[5] indicate a common basic mechanism - strong hybridization - for the Ce, Pr, Nd, and Pm instabilities. The drastic variations of the f count, spin and degeneracy factors, of the strength of hybridization as a function of the matrix and of the 4f ion species, and in particular of the locations of the 4f levels, might all be essential reasons to extend theoretical studies from Ce to Pr, Nd, and Pm systems.

Probably, the 4f shell of Ce in W or Re is the most strongly hybridized of the metallic 4f systems. This estimate is based on the extrapolation of the degree of instability found for Pm, Nd, and Pr to Ce in W and Re hosts[4]. For Ce in W and Re the atomic correlation energies might be smaller than the linewidth, so that the 4f shell can be delocalized.

NEARLY LOCALIZED 3d AND 4d STATES IN METALS

How can one produce more localized 3d systems and magnetic 4d systems? The leading term for localization tendencies seems to be the hybridization, the role of the density of states at E_F might be less decisive. One way to decrease hybridization is implantation of d ions into the large-volume alkali metal hosts[2].

With Fe[6], Mo[7], Tc, and Ru[8], recoil implanted into alkali metal hosts, it can be possible to find d systems, in which the atomic correlations are much larger than the d linewidths. For Fe, Tc and Ru in Rb and Cs hosts, we have found large orbital contributions to the magnetic hyperfine fields and to the moments, which are about an order of magnitude larger compared to those found in some alloying 3d systems like Co in Au[9]. It has also been possible to observe the spin dynamics for Fe[6] and Mo[7] in alkali metals, which comes out to be extremely small. Large orbital contributions and small spin rates, are both indicative for strong localization tendencies. Fe in alkali metals behave like a nearly stable 4f system, described by a $3d^6$ ground state of Fe^{2+} with L = 2, S = 2, J = 4 in LS coupling.

As a crucial mechanism for the magnetic behaviour of certain d ions in Rb and Cs a drastic decrease of hybridization is suggested, mainly due to the large differences in volume of the d cells and the alkali cells, and due to only s-like character of the host conduction electrons.

The finding of nearly localized 3d and 4d states in metals implies a number of consequences and new problems. The data permit a comparison of magnetic 3d with magnetic 4d systems, of 3d and 4d with 4f systems. They also permit a comparison of the magnetism of e.g. Fe in alkalis to Fe in other host metals. It is possible to deduce information of the 3d and 4d ground states of these new atoms in metals.

Possible crystal field effects for 3d and 4d ions in alkali metals are surprisingly weak. The total splitting for Fe in Rb and Cs comes out to be smaller than 0.025 eV [6] and for Tc and Ru in Rb and Cs the splitting is comparable or smaller than the LS coupling. These findings suggest a more critical view of the assumption that crystal field splittings are large compared to the temperatures at which 3d ions in metal hosts were measured. In my opinion we have to reinvestigate the role of crystal field effects on the local 3d magnetism in general.

COMPARISON OF 3d- AND 4d- WITH 4f INSTABILITIES

The observation of localized 3d and 4d shell behaviour allows a comparison of magnetic d systems with 4f systems on a more reliable basis than before.

According to the spin rates observed, the systems Mo and Fe in Rb and in Cs are among the most stable d systems known hitherto (see Fig. 1). In selected metallic systems, the 3d and 4d hybridization seems to be comparable or even smaller than the 4f mixing. On the other hand, the spin rates for Fe and Mo in alkalis are large compared to Gd systems (Fig. 1), so that one can characterize Fe and Mo in Rb and Cs as Kondo systems with very small T_K values.

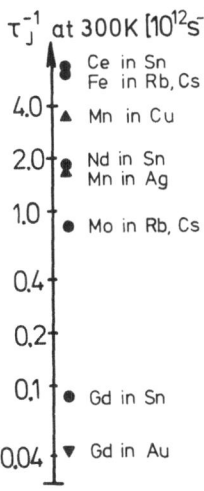

Fig. 1. Spin rates τ^{-1}_J in selected systems as observed by ESR (7) neutron scattering (8) and TDPAD (9), all are extrapolated to 300 K.

The instabilities observed for e.g. Fe in Li [6] and in Ca [10] have much in common with the 4f instabilities of Ce, Pr, Nd, and Pm ions under high lattice pressure. The 3d and 4d instabilities in alkali and alkaline earth metals might be dominated by strong hybridization, so that the d linewidth is larger than a possible crystal field splitting. An analogous interpretation has been given for Ce, Pr, Nd, Pm instabilities as a function of increasing lattice pressure[4].

As far as I can see, such an analogy to 4f instabilities is only valid for 3d and 4d ions in s metals and perhaps in sp metals. For 3d ions in d hosts and in Cu, Ag, Au, such a simple analogy to 4f systems does not exist.

ON THE ORIGIN OF LOCAL Fe MAGNETISM IN METALS

The local Fe moment in nearly all metallic systems has been parametrized by an effective spin S_{eff}. Very often the 3d electrons are assumed to be itinerant and orbital contributions are assumed to be quenched by large hybridization and/or crystal fields. Why and under which conditions can one expect the ionic-type magnetism (e.g. Fe in Cs) and the S_{eff}-type magnetism (e.g. Fe in Mo, Pd, Au, Fe)?

The data obtained by dedicated TDPAD experiments and data known in the literature are consistent with the following results[11]: (i) The existence of d band electrons in the host plays a crucial role for the type of Fe magnetism observed. S_{eff}-type magnetism occurs in hosts with d electrons whereas for Fe in s and sp metals positive hyperfine fields or nearly non-magnetic behaviour is observed (with perhaps one exception). (ii) This is also valid for Fe in Cu, Ag, and Au. The Fe moments in the noble metals are predominantly determined by the interaction of the Fe d with the 3d, 4d and 5d host electrons, respectively. Under increasing hybridization (increasing lattice pressure) Fe in s and sp metals changed from ionic-type to nonmagnetic behaviour; similar to the behaviour of certain ions in metals.

These findings suggest that the Fe spin magnetism is strongly correlated to interactions of the impurity d with at least the d electrons of its nearest neighbours. Strong Mn d-host d hybridization has also been suggested for Mn in noble metals, based on electron spectroscopies[12]. One is led to argue that strong impurity 3d-host d hybridization has important influence on spin exchange splittings, on crystal field splittings and perhaps on the spin linewidths. Furthermore, there could be a strong influence on collective interactions between magnetic d ions.

ACKNOWLEDGEMENTS

It is a pleasure to acknowledge the hospitality and generous support of the Hahn-Meitner-Institut, in particular the support given by Prof. Dr. K.H. Lindenberger. This work was supported by the Deutsche Forschungsgemeinschaft and by the Bundesminister fur Forschung und Technology. I would like to acknowledge discussions with many colleagues, in particular with those in Berlin and in Cologne.

REFERENCES

1.   P.W. Anderson, Phys. Rev. 124:41 (1961).
     L.L. Hirst, Phys. Kondens. Mater. 11:255 (1970), and Z. Phys. 214:9, 278 (1971).
2.   D. Riegel, J. Magn. Magn. Mater. 52:96 (1985).

3. H.J. Barth, M. Luszik-Bhadra, and D. Riegel, Phys. Rev. Lett. 50:608 (1983).
4. L. Buermann, H.J. Barth, K.H. Biedermann, M. Luszik-Bhadra, and D. Riegel, Phys. Rev. Lett. 56:492 (1986).
5. D. Riegel, Phys. Rev. Lett. 48:516 (1982).
6. D. Riegel, H.J. Barth, L. Buermann, H. Haas, and Ch. Stenzel, Phys. Rev. Lett. 57:388 (1986).
7. K.D. Gross, M. Luszik-Bhadra, and D. Riegel, Phys. Rev. Lett., submitted.
8. K.D. Gross, Th. Kornrumpf, and D. Riegel, to be published.
9. R. Dupree, R.E. Walstedt, and W.W. Warren, Jr., Phys. Rev. Lett. 38:612 (1977).
10. M.H. Rafailovich, E. Dafani, H.E. Mahnke, and G.D. Sprouse, Phys. Rev. Lett. 50:1001 (1983).
11. D. Riegel, L. Buermann, K.D. Gross, and Th. Kornrumpf, to be published.
12. D. van der Marel, C. Westra, G.A. Sawatzky, and F.U. Hillebrecht, Phys. Rev. B31:1936 (1985).

POINT-CONTACT SPECTROSCOPY IN A HEAVY-FERMION METAL: UPt_3 vs Cu

A.G.M. Jansen, A.M. Duif, A.A. Lysykh[*] and P. Wyder

Max-Planck-Institut für Festkörperforschung
Hochfeld-Magnetlabor
25, Avenue des Martyrs, 166X, F-38042 Grenoble Cedex, France

[*]Permanent address: Institute for Low Temperature Physics
and Engineering, Ukranian S.S.R. Academy of Sciences
47 Lenin Avenue, Kharkov, USSR

ABSTRACT

For point contacts with the heavy-fermion system UPt_3 the non-linearities in the current-voltage characteristics are investigated as a function of the bath temperature. The data are discussed in terms of local heating of the contact area. For comparison, point contacts of Cu show energy-resolved spectroscopy of the electron scattering.

In recent years various experiments have been done on metallic point contacts with anomalous rare-earths and actinides[1][7]. With point contacts between more simple metals the measurement of the nonlinear current-voltage characteristics had proved to be useful as a spectroscopic method to study the interaction of the conduction electrons with elementary excitations in a metal (phonons, magnons, paramagnetic impurities, crystal-field levels)[8,9]. As a matter of course it was very interesting to apply the point-contact method to the above-mentioned exotic compounds, where the partly filled f-shell gives rise to phenomena like valence fluctuations and heavy fermions. In these systems the observed non-linear structure in the point-contact characteristics was explained in terms of the inelastic scattering with valence fluctuations[1] and the energy-dependent density of states[3][5]. However, alternative explanations were given in terms of local heating of the contact area due to the strong scattering of the electronic system[6,7]. To clarify this problem, we will describe the point-contact method by mentioning the important parameters (contact diameter with respect to electron mean free path) and their influence on the measured spectra. Experimental results will be presented for the heavy-fermion system UPt_3 and the simple noble metal Cu to show the difference between these metals in doing point-contact spectroscopy.

The dimension of a contact can be defined as the radius a of a circular contact between two semi-infinite metallic parts. Another parameter in point-contact spectroscopy is the mean free path $l(\varepsilon)$ of the electrons for inelastic scattering. This mean free path depends on the electron energy ε with respect to the Fermi level. For the case of a clean contact $(a \ll l(\varepsilon))$, the electrons will be accelerated within a mean free path upon passing the constriction at an applied voltage V. Then, the relevant mean free path involves the electron energy ε = eV. Sharvin[10] realized the importance of a metallic contact in the clean limit and estimated the resistance R_{Sh} = $4\rho l/3\pi a^2$ for a circular contact with electrical resistivity ρ of the bulk material. The ballistic injection of the hot electrons results in a non-equilibrium distribution of the electrons near the contact: the filled Fermi sphere is distorted with differences in the kinetic energy depending on the velocity direction (electrons coming from the low or high potential side of the contact). Relaxation of the accelerated electrons may occur via the interaction with excitations in the metal, for instance by spontaneous emission of phonons. In these scattering processes the electrons can flow back through the contact yielding voltage-dependent corrections of the current. In the ballistic limit ($a \ll l(\varepsilon)$) of point-contact spectroscopy the main result[8,9] for the interpretation of the observed non-linearities can be summarized by the following proportionality

$$d^2I/dV^2 \propto -a^3 G(eV). \tag{1}$$

The second derivative d^2I/dV^2 of the current with respect to the voltage is proportional to the energy-dependent spectral function G for the scattering of the electrons. For the electron-phonon interaction the function G equals the well-known Eliashberg function α^2F with a slight modification for the kinematic efficiency in the back-flow processes. The signal intensity of Eq. 1 is proportional to the contact volume a^3 where the back-flow collisions contribute most efficiently. In Fig. 1 we have plotted the point-contact spectra (dV/dI and d^2V/dI^2) for a point contact of Cu at helium

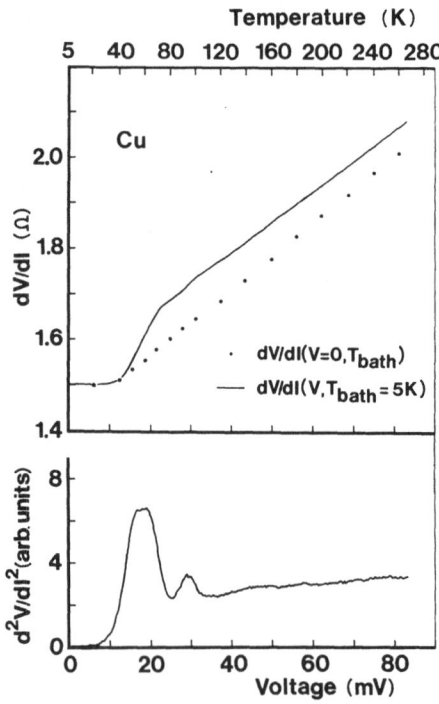

Fig. 1.   The dynamical resistance dV/dI and the second derivative d^2V/dI^2 for a Cu-Cu point contact at 5 K as a function of the applied voltage, revealing the two-peak structure of the transverse and longitudinal phonons. For the same contact the resistance at zero voltage has been plotted as a function of the bath temperature with a relative scaling (3.2 K/mV) given by the heating model in Eq. 2 with the Sommerfeld value of the Lorenz number.

temperatures. The measured second derivative $d^2V/dI^2 \approx -(d^2I/dV^2)(dV/dI)^3$ clearly shows the phonon structure due to transverse and longitudinal phonons. The point-contact method gives the possibility in normal metals to study the energy dependence in the strength of the electron-phonon interaction by comparing the point-contact spectra with the phonon density of states F from inelastic neutron-scattering data.

For less pure materials (elastic mean free path $l_i < a$) it is still possible to do point-contact spectroscopy provided that the inelastic diffusion length $\Lambda(\varepsilon) = [l_i l(\varepsilon)]^{1/2}$ will be large compared to the contact radius. The obtained spectral information is described analogously to Eq. 1. The efficiency volume is now given by $a^2 l_i$. Experimentally, this results in a reduction of the observed signals by a factor l_i/a.[11]

In the discussed ballistic and diffusive limit the characteristic length for inelastic scattering is large compared to the contact dimension. In first order heating effects can be neglected because most of the phonons will be generated far from the contact. In a more detailed look at the d^2V/dI^2-spectrum in Fig. 1 it is seen that the signal does not vanish for voltages above the corresponding Debye energy (29.5 meV for Cu). The observed background of the signal is due to scattering of the electrons with the non-equilibrium phonons (stimulated emission and absorption) generated in the contact area. In the case of strong inelastic scattering heating effects play a dominant role.

In the thermal limit ($l(\varepsilon), \Lambda(\varepsilon) \ll a$) energy resolution of the electron scattering is not possible. The electron and phonon system are the thermal equilibrium and the Joule heat will locally heat up the contact area. Under the assumption of equal flow lines for the electrical and thermal conduction one finds a relation between the maximal temperature T_{max} at the center of the contact and the applied voltage[12]

$$V^2 = 8 \int_{T_{bath}}^{T_{max}} \rho\lambda \, dT, \qquad (2)$$

for a contact at a temperature T_{bath} with an arbitrary geometry between metals with a temperature-dependent resistivity ρ and thermal conductivity λ. If the Wiedemann-Franz law $\rho\lambda = LT$ holds, this expression yields for $T_{max} \gg T_{bath}$ a temperature increase of 3.2 K for each mV across the contact with the Sommerfeld value of the Lorenz number $L = 2.45 \ 10^{-8} \ V^2/K^2$. In the dirty limit the point-contact resistance R_M is proportional to the bulk resistivity ($R_M = \rho/2a$)[12]. Therefore, the voltage dependence of the contact resistance will resemble the temperature dependence of the bulk resistivity using Eq. 2 for the mapping between the applied voltage V and the local contact temperature T_{max}.

For a resistance R_{Sh} of 1 Ω, in the clean limit, we get for metals with a typical value $\rho l = 10^{-15} \ \Omega m^2$ a contact dimension a = 20 nm. Assuming that the mass renormalization has no effect on the parameter ρl, this estimate holds for both the noble metal Cu and the heavy-fermion system UPt_3. For pure Cu, the electron mean free path will be larger at low temperatures and we can investigate a contact in the clean limit. For UPt_3, the value $\rho(4 \ K) = 3 \ 10^{-7} \ \Omega m$ [13] yields a mean free path of the order of 3 nm and the contact is not in the clean limit. To estimate the contact dimension for a resistance of 1 Ω, we have to use the expression for R_M, yielding a = 150 nm. Note that for energy-resolved spectroscopy we have to consider the energy-dependent mean free path $l(\varepsilon = eV)$ at low temperatures. For the electron-phonon interaction in Cu this value is minimal at the Debye energy ($\approx 1 \ \mu m$). In UPt_3 this mean free path will be dominated by strong electron-electron scattering with a much smaller value.

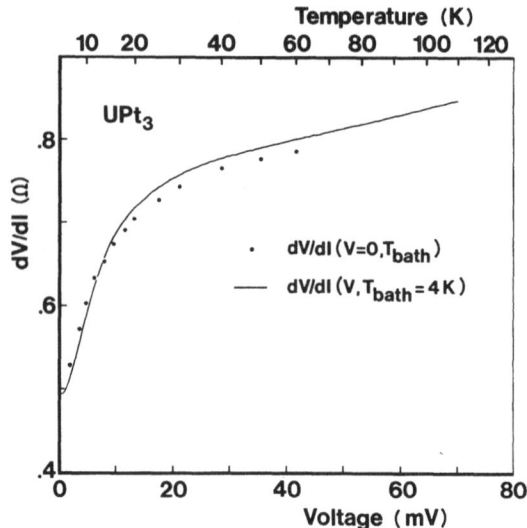

Fig. 2.   The dynamical resistance dV/dI(V,T $_{bath}$ = 4 K) as a function of
the voltage at low temperatures and the resistance dV/dI(V=0,T$_{bath}$)
as a function of the bath temperature at zero bias for the same
UPt$_3$-UPt$_3$ point contact. The scaling of the temperature with
respect to the voltage ($\approx$ 1.5 K/mV) has been obtained from the
heating model in Eq. 2 using experimental data for the resistivity
and thermal conductivity.[14]

In Fig. 2 we have plotted the dynamical resistance dV/dI of a UPt$_3$
point contact as a function of the applied voltage, measured at 4 K. The
resistance shows a sharp drop at voltages below 20 mV. The functional dependence dV/dI(V) resembles the temperature-dependent bulk resistivity ρ(T)
of UPt$_3$.[13] Spin-fluctuation scattering (ρ(T) $\propto$ T^2) and a Kondo-lattice
formation may cause the drop of the resistivity at low temperatures. For
the point contact it is better to use the temperature-dependent contact
resistance as a measure for the local resistivity in the constriction. In
Fig. 2 we added for the same contact the measured data for the temperature
dependence of the contact resistance at zero voltage. The adjustment between
the voltage and temperature scale was done using the relation between V
and T$_{max}$ in Eq. 2 with experimental data for the electrical resistivity
and thermal conductivity of UPt$_3$.[14] By plotting the data in this way, Fig.
2 shows the agreement in functional dependence of dV/dI(V) and dV/dI(T).

The temperature dependence of the point-contact spectra has been studied
for Cu in detail[15]. For the data of Cu in Fig. 1 we have also plotted the
temperature-dependent contact resistance for a comparison with the voltage-
dependent resistance. The adjustment between voltage and temperature scale
has been done with the Sommerfeld value for the Lorenz number in Eq. 2.
For the Cu point contact we see a difference between the data of
dV/dI(V = 0,T$_{bath}$) and dV/dI(V,T$_{bath}$ = 5 K). The phonon features in the
voltage-dependent spectra are not revealed in the temperature-dependent
contact resistance. A description in terms of the heating model fails for
pure contacts.

The energy dependence of the electron density of states drops out in
the description of a clean contact for a simple metal with parabolic bands,
because it is cancelled by the electron velocity in the expression for the
current. A detailed consideration of non-parabolic bands, anisotropic and
mass-renormalization can still have an effect on the voltage-dependent
contact resistance, related to the heavy-fermion character. However, a

theory to incorporate this is not present. Phonons are not observed in the measured point-contact spectra of UPt_3, probably due to the strong inelastic electron-electron scattering in a heavy-fermion metal. In view of this strong inelastic scattering, one must be careful in the analysis of point-contact data in terms of a ballistic model with a long mean free path $l(eV)$ of the electrons compared to the contact dimension. The temperature-dependent measurements for UPt_3 indicate that thermal effects with local heating of the contact area are important.

REFERENCES

1. B. Bussian, I. Frankowski, and D. Wohlleben, Phys. Rev. Lett. 49:1026 (1982).
2. E. Paulus and G. Voss, J. Magn. Magn. Mat. 47&48:539 (1985).
3. I. Frankowski and P. Wachter, Solid State Commun. 41:577 (1982).
4. M. Moser, P. Wachter, F. Hulliger, and J.R. Etourneau, Solid State Commun. 54:241 (1985).
5. M. Moser, P. Wachter, and J.J.M. Franse, Solid State Commun. 58:515 (1986).
6. Yu. G. Neidyuk, N.N. Gribov, A.A. Lysykh, I.K. Yanson, N.B. Brandt, and V.V. Moschalkov, Pis'ma Zh. Eksp. Teor. Fiz. 41:325 (1985) [JETP Lett. 41:399 (1985).
7. A.G.M. Jansen, A. de Visser, A.M. Duif, J.J.M. Franse, and J.A.A.J. Perenboom, J. Magn. Magn. Mat. 63&64:670 (1987).
8. A.G.M. Jansen, A.P. van Gelder, and P. Wyder, J. Phys. C13:6073 (1980).
9. I.K. Yanson, Fiz. Nizk. Temp. 9:676 (1983) [Sov. J. Low Temp. Phys. 9:343 (1983).
10. Yu.V. Sharvin, Zh. Eksp. Teor. Fiz. 48:984 (1965). [Sov. Phys. JETP 21:655 (1965)]
11. A.A. Lysykh, I.K. Yanson, O.I. Shklyarevski and Yu. G. Naydyuk, Solid State Commun. 35:987 (1980).
12. R. Holm, in Electric Contacts (Springer Verlag, Berlin, 1967).
13. A. de Visser, J.J.M. Franse, and A. Menovsky, J. Magn. Magn. Mat. 43:43 (1984).
14. J.J.M. Franse, A. Menovsky, A. de Visser, C.D. Bredl, U. Gottwick, W. Lieke, H.M. Mayer, U. Rauchschwalbe, G. Sparn, and F. Steglich, Z. Phys. B59:15 (1985).
15. A.P. van Gelder, A.G.M. Jansen, and P. Wyder, Phys. Rev. B22:1515 (1980).

"SOFT"- VERSUS "HARD" SPECTROSCOPIES FOR NARROW BAND MATERIALS

Peter Wachter

Laboratorium für Festkörperphysik, ETH Zürich
8093 Zürich, Switzerland

INTRODUCTION

The knowledge of the electronic structure of intermediate valence - and heavy fermion materials is of fundamental importance to understand the thermodynamic behavior, the electrical resistivity or the magnetic properties. The quoted physical entities are heavily dependent on the electronic structure within meV around the Fermi energy E_F, especially when the electronic density of states is a strongly varying function of the energy near E_F. Present day band structure calculations are not yet able to give us an energy resolution of the order of meV[1]. Besides there is no simple way how to include the many body interactions of the electrons near E_F into the band structure calculation, so various ad hoc enhancement factors have do be proposed to match the calculated density of states at E_F with e.g. the measured γ value of the specific heat.

This unsatisfactory state of the art can only be overcome by trying to measure the real existing electronic structure near E_F, including all many body interactions, and to try to derive from these measurements a quasiparticle band structure. However, there is no simple answer as to what the various experimental methods measure.

Intermediate valence and heavy fermions have so far been only detected on 4f- or 5f compounds or alloys, respectively. It is generally accepted that the hybridization of the f states with a wide band, usually a d band, plays a dominant role. This hybridization has been taken into account in the single impurity Anderson Hamiltonian[2], but it has been shown theoretically[3,4,5] and experimentally[6,7] that the lattice case and not a single f like impurity has to be considered. The theoretical treatment then invariably results in a two peak structure of f like character which, together with a more or less classical d band, is shown in Fig.1 for UPt_3 and $CeCu_2Si_2$. The particles in these narrow f like bands are duely termed quasiparticles with a heavy mass where a many-body enhancement of the mass is thought to arise because of the interactions of the f electrons.

Experimentally the question arises whether these quasiparticle bands are realistic, how they can be measured and how they compare with f bands derived from band structure calculations. We think that in the present band structure calculations of e.g. UPt_3[1] not all electron interactions are included, thus the band effective mass will not be the same as the quasiparticle mass. In fact, mass enhancement factors of more than 20 had to be assumed to match the experimentally observed γ values. Recently de Haas-van Alphen measurements have been performed on $CeCu_6$[8] and UPt_3[9] and heavy effective masses have been determined. The main result from this experiment is that Fermi surfaces and quasiparticle bands exist. For the experimental determination of these quasiparticle bands it is important to know how much energy is transfered to the quasiparticle during a e.g. spectroscopic measurement. A border line is constituted by the interaction energy of f-and d electrons, which is the hybridization energy with a typical magnitude of 10 meV[10]. Thus excitation energies larger than about 10 meV will destroy the heavy quasiparticle state and eject one undressed electron as in UPS or XPS experiments. The position in energy of this bare electron state (in Ce compounds about 2 eV below E_F) which does not exist in the solid before the emission process, because f electrons in the solid have many body interactions, has often been taken as the position of the ground state of the f electrons. Whether one can reconstruct the initial quasiparticle state as a Kondo resonance from this measured final state spectrum using the Gunnarson-Schönhammer mechanism[11] remains questionable since this theory uses the single impurity Anderson Hamiltonian[2] and not the lattice case and as a consequence cannot reproduce the two-peak electronic structure depicted in Fig. 1.

It is then important to use as experimental methods for the investigation of the quasiparticle state spectroscopies which transmit less than about 10 meV to the dressed electrons and those are e.g. far-infrared- (FIR) and point contact spectroscopy (PCS). In addition these methods must be used at very low temperatures because kT at 300 K corresponds already to 25 meV. We thus have applied these spectroscopies to intermediate valent materials like SmB_6, "gold" SmS, YbB_{12} and TmSe and to heavy fermions like UPt_3, $CeCu_6$, and with PCS alone to $CeAl_3$, U_2PtC_2, U_6Co and UIr_2Si_2. It will be shown that the electronic structure as depicted in Fig. 1 holds as well for intermediate valent- as for heavy fermion materials, the difference being the position of the Fermi energy in this structure.

EXPERIMENT

PCS measures the band width of the narrow quasiparticle band in which the Fermi energy lies[12] and it turns out to be in the meV range for the various heavy fermion materials. With FIR we have observed for the first time on UPt_3[7] and now on $CeCu_6$ that there exist interband transitions with f character also in the meV range. Thus we have to assume at least two quasiparticle bands, between which an optical transition is possible (Fig. 1). In the case of the uranium compounds the lowest band is completely full and the upper partially occupied, in the case of the cerium compounds, the lower band is partially full and the upper one is empty. The partially filled band is responsible for the high γ value and it carries the quasiparticles with the high effective mass as shown by the de Haas-van Alphen effect[8,9]. The important and new observation is that we can have thermally and optically induced transitions between the two quasiparticle bands and that at temperatures above T=0 both bands are carrying electrical current. In ref.[7] we could determine from the optical plasma resonance of the quasiparticels in combination with the γ value that for UPt_3 the effective mass $m^* = 250$ m and that the concentration of carriers in the quasiparticle band at E_F is one per uranium and that this band is exactly half full. Since in total we have to accomodate three 5f electrons per uranium we must conclude that in the other f like quasiparticle band two particles per uranium fill this band completely. In the case of UPt_3 a pseudogap Δ of 4 meV (6 meV for $CeCu_6$) has been measured by FIR (defined as the separation of the two density of states peaks in Fig. 1) and a band width BW of the band at E_F of ~ ~ been determined by PCS (2 meV for CeCu$_6$)[12].

THE MODEL

When we now assume that for heavy fermions we have an electronic structure consisting of at least two narrow quasiparticle bands, each one capable of holding two particles per f atom we have in principle a band structure as in ref.[1] where the many body interaction not included in ref.[1] is expressed in the larger effective mass. Experimentally it is difficult to say that one has more than two quasiparticle bands, although there may be evidence for this[7], but for the determination of the thermodynamic properties only the bands near E_F are of importance. Such a two-band structure has also been obtained in the f-d hybridization model of Brandow[4]. When we adopt now such a rigid density of states model we find that the main influence of temperature is the change of the Fermi distribution function, which at room temperature is much larger than the width of the whole electronic structure near E_F.

With this in mind we have calculated in ref.[13,14] the temperature dependence of the electrical resistivity of UPt$_3$, U$_2$PtC$_2$, CeCu$_2$Si$_2$, CeCu$_6$, CeAl$_3$ and UIr$_2$Si$_2$, the γ value of UPt$_3$, CeCu$_2$Si$_2$ and of CeCu$_6$ and the magnetic susceptibility of UPt$_3$ and CeCu$_6$[15] assuming a Gaussian distribution of the density of states as depicted in Fig. 1, which corresponds to two narrow quasiparticle bands hybridized with a wide d band. Excellent agreement with the measurements of the various authors and the experimental values determined by FIR and PCS is obtained.

CONCLUSION

We have shown that the quasiparticle band structure can indeed be measured when using "soft" or low energy spectroscopies which transmit so little energy to the quasiparticle that it is not shaken out of its interactions with other quasiparticles. Such a band structure is in principle calculated by refs.[1,4] and confirmation of such quasiparticle bands with heavy masses comes from the de Haas-van Alphen effect[9], also a "soft" spectroscopy. The magnetic fields of 10 T used in these experiments, however, change the band structure near E_F significantly, so that the value of the effective mass of 90 m remains questionable.

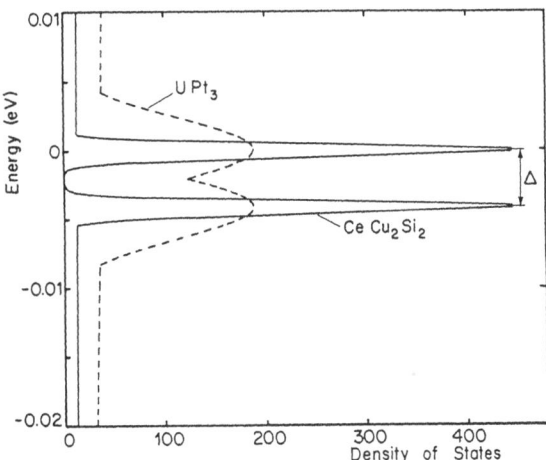

Fig. 1: Quasiparticle band structure of UPt$_3$ and CeCu$_2$Si$_2$.

In contrast to the rigid density of states structure of refs.[1-4] the Kondo resonance disappears above the Kondo temperature. However, it has been shown experimentally with PCS[12] that narrow quasiparticle bands exist still high above the Kondo temperature. Besides, in the Kondo model the spectral weight of the Kondo resonance is very low and it should not influence the dielectric constants. The contrary has been observed experimentally: the dielectric constants caused by the narrow quasiparticle band structure amount to 20'000[7].

REFERENCES

1.   T. Oguchi and A.J. Freeman, J. Magn. Magn. Mat. 52: 174 (1985)
2.   P.W. Anderson, Phys. Rev. 124: 41 (1961)
3.   N. Grewe, Solid State Commun. 50: 19 (1984)
4.   B.H. Brandow, Phys. Rev. B 33: 215 (1986)
5.   C. Lacroix, J. Magn. Magn. Mat. 63&64: 239 (1987)
6.   G. Travaglini and P. Wachter, Phys. Rev. B 29: 893 (1984)
7.   F. Marabelli, G. Travaglini and P. Wachter, Solid State Commun. 59: 381 (1986)
8.   P.H.P. Reinders, M. Springford, P.T. Coleridge, R. Boulet and D. Ravot, Phys. Rev. Lett. 57: 1631 (1986)
9.   L. Taillefer, R. Newbury, G.G. Lonzarich, Z. Fisk and J.L. Smith, J. Magn. Magn. Mat. 63&64: 372 (1987)
10. C.M. Varma, Rev. Mod. Phys. 48: 219 (1976)
11. O. Gunnarson and K. Schönhammer, Phys. Rev, B 28: 4315 (1983)
12. M. Moser, P. Wachter, J.M.M. Franse, G.P. Meisner and E. Walker, J. Magn. Magn. Mat. 54-57: 373 (1986)
13. F. Marabelli, P. Wachter and J.M.M. Franse, J. Magn. Magn. Mat. 62: 287 (1986)
14. F. Marabelli and P. Wachter, Proc. Int. Conf. Valence Fluctuations, ICVF, Bangalore, India, (1987)
15. F. Marabelli and P. Wachter, Int. Symp. Magnetism of Intermetallic Compounds, ISMIC, Kyoto, Japan, (1987)

B

THEORETICAL DESCRIPTIONS OF THE GROUND STATE

SECTION B: THEORETICAL DESCRIPTIONS OF THE GROUND STATE

INTRODUCTION

The search for good theoretical descriptions of the electronic structure
and related physical properties of narrow band materials started at least
50 years ago and is still the primary occupation of a large group of chemists
and physicists. Although a lot of progress has been made one must admit
that we are still far removed from a general theoretical description of
narrow band materials. Rather than being able to predict heavy fermion
behaviour in say UPt$_3$ or superconductivity in non-stoichiometric oxides, the
theoretical community is continually caught off guard by such experimental
discoveries. Although "predictions" after the fact are quite common and
often successful, the situation of the high T$_c$ materials seems to be a lot
different. As Falicov states at this workshop "there are now N theories
for high T$_c$ materials (N $\simeq$ 20) of which at <u>least</u> N-1 are not applicable".

The theoretical problem we are faced with is to find a suitable approximation for describing a system of a large number of interacting electrons
in a (periodic) potential of ion cores. Basically two types of approaches
have been developed, namely the ab initio approaches without adjustable
parameters and the model Hamiltonian approaches with empirically determined
parameters. In the ab initio approaches the approximation is either to
limit the number of electrons by replacing the solid by a small cluster or
by replacing the effect of the electron-electron interaction by an effective
one particle potential as in density functional band theory. In the model
Hamiltonian approaches the main effort is to obtain solutions to the many
body problem but with highly simplified interactions in an attempt to qualitatively describe the physical properties of narrow band materials. All of
these approaches have contributed greatly to our present day understanding
of narrow band materials but each of these approaches has its successes
and failures both of which have been openly discussed at this conference.
The cluster theory methods, apparently because of their neglect of long
range screening effects, obtain much too large band gaps and, because of
their neglect of translational symmetry, cannot describe the types of magnetic order or the conductivity. They are, however, quite successful in describing local excitations involving multiplet structures and crystal and
ligand field splittings.

Density functional methods are very successful in describing the cohesive energy, charge density distribution, type of magnetic order and
Fermi surfaces of many solids. They however often predict much too small
band gaps, metallic behaviour for known insulators and are not able to

describe the temperature dependent properties.

The model Hamiltonian approaches have been successful in describing the concepts of correlation gaps and the physics of other many body effects like the Kondo problem, mechanisms for exchange and superexchange interactions as well as qualitative or semi-empirical descriptions of thermal properties. These methods are, however, parameter dependent and often completely different models can result in similar predictions as just mentioned above for the high T_c materials.

The starting point of a thorough theoretical discussion of narrow band materials could be taken as 1929 with Bloch's theorem[1] closely followed by Wilson's[2] theory with which crystalline materials could be classified into metals and insulators according to whether or not their electronic band structures involved any partially filled Bloch bands. It did not take too long, however, before the first exceptions to these rules were pointed out by de Boer and Verwey[3] who showed that many 3d transition metal compounds (oxides) are insulators whereas Wilson's scheme would have predicted them to be metals. In the discussion of that paper Peierls pointed out that for NiO the repulsive Ni d-d Coulomb interaction is larger than the kinetic energy gained by delocalization and therefore each Ni ion would have 8 d electrons and only spin degrees of freedom would remain. These ideas were formulated by Mott[4,2] who argued that the insulating behaviour of many transition metal compounds was due to a large d-d Coulomb interaction (U) which supresses polarity fluctuations and can cause a correlation gap to open for conduction. Some years later Hubbard[6] introduced a model Hamiltonian which includes the possible competition of kinetic energy lowering by banding and Coulomb correlation energy lowering by localization. Hubbard argued that for tight binding like narrow band systems the Coulomb interaction could to lowest order approximation be replaced by the on site interaction U resulting in the much studied Hamiltonian for an s band of the form

$$H = \sum_{k\sigma} \varepsilon_{k\sigma} c^+_{k\sigma} c_{k\sigma} + \sum_{i\sigma} \frac{U}{2} c^+_{i\sigma} c_{i\sigma} c^+_{i\sigma} c_{i\sigma}$$

in which $\varepsilon_{k\sigma}$ describes the banding and U is defined by

$$U = E(S^2) - E(S^1) + E(S^0) - E(S^1) = E_A + E_I$$

where $E_{A(I)}$ is the electron affinity (ionization potential) of a singly occupied ion screened by the polarizability of the solid. The exact value of U is therefore dependent not only on the atom or ion but also on the solid and its frequency dependent dielectric constant. Hubbard has shown that such a model Hamiltonian yields an insulating material for a half-filled band for $U > W$ where W is the one electron dispersional band width, and a metal for $U < W$. Also for $U \geq W$ the only low energy scale degrees of freedom remaining are those due to the spin and these materials are in general magnetically ordered. The low energy scale physics can be described with spin only Heisenberg like Hamiltonians.

Just before this Anderson[7] had used a very similar approach to describe the interatomic exchange interaction appearing in the Heisenberg spin Hamiltonians in insulating transition metal compounds. In this case U refers to the Coulomb interaction of d electrons and is defined by

$$U = E(d^{n+1}) - E(d^n) + E(d^{n-1}) - E(d^n)$$

Because of their insulating character, Anderson took U >> d band width and showed that interatomic exchange interactions are due to virtual charge fluctuations involving U but mediated by intervening anions.

Such a superexchange interaction involves the charge transfer energy which Anderson implicitly assumed to be large compared to U, a point which is now being questioned[8]. The Mott-Hubbard model together with Anderson's theory of superexchange has had a tremendous impact on transition metal and rare earth compound research. Together with the van Vleck-Orgal theory of ligand fields[9][10], the Sugano Tanabe diagrams[11], and the Goodenough Kanamori rules[12] for superexchange, one was able to explain much of the systematics in the magnetic, optical and electrical properties of transition metal and rare earth compounds.

Another important area related to narrow band phenomena, which was developing in parallel, involved the disruption of magnetic impurities in metals and the problem of magnetic moment formation. Friedel[13] and Anderson [14] played the leading roles in this area. In this connection Anderson introduced a model Hamiltonian which has formed a basis for describing magnetic impurities in metals. This Hamiltonian was based on Friedel's[13] suggestion that for 3d transition metal impurities the d orbitals would retain their localized character even if dissolved in a metal, although the states would be broadened and shifted, and are referred to as virtual bound states. Friedel also suggested that Hund's rule would then tend to split the minority and majority spin states and a magnetic moment would result if this splitting was large compared to the width of the state. Anderson followed this up and, as in the Hubbard model, neglected all Coulomb interactions except that between two electrons in the impurity d(f) shell.

$$H = \sum_{k\sigma} \varepsilon_k c^+_{n\sigma} c_{k\sigma} + \sum_m \varepsilon_{dm} d^+_m d_m + \sum_{ijlm} U(ijlm) d^+_i d_j d^+_l d_m +$$

$$+ \sum_{km} V_{km} [d^+_m c_{k_m} + c^+_{k_m} d_m]$$

The first term in this Hamiltonian describes the host metal valence band structure, the second and third - the impurity atomic structure including the multiplet structure and the last term - the hybridization of the impurity states with the valence bands of the host. Although simple in form, this Hamiltonian turns out to reveal a surprising richness in physical phenomena and is remarkably difficult to solve. The basic electronic structure depicted by this model Hamiltonian is shown in Fig. 1. The ground state is assumed to have n localized d or f electrons on the impurity so that the states below involve d^{n-1} states and those above d^{n+1} states. The minimum separation of these states is U. For both $E(d^{n-1})$ and $E(d^{n+1})$ much larger than the virutal bound state broadening $\Gamma = \pi_\rho |v_{km}|^2$ (which requires U to be large also) a localised moment exists and, as shown by Schrieffer and Wolff[15], this Anderson Hamiltonian can be transformed to a Kondo Hamiltonian[16] in which the impurity retains only its spin degrees of freedom

$$H = \sum_{k\varepsilon} \varepsilon_k c^+_{k\sigma} c_{k\sigma} - \frac{J}{2N} \sum_{kk}{}' [(c^+_{k\uparrow} c_{k'\downarrow} - c^+_{k\downarrow} c_{k'\downarrow}) S_z$$

$$+ c^+_{k\uparrow} c_{k'\downarrow} S_- + c^+_{k\downarrow} c_{k'\uparrow} S_+]$$

and S is the impurity spin operator. Also the Hamiltonian is extremely difficult to solve, although renormalization group methods have led to at least partied solutions[17].

Even for U large however we could still imagine a situation where either $E(d^{n-1})$ or $E(d^{n+1})$ is small. For example, by putting the same impurity into

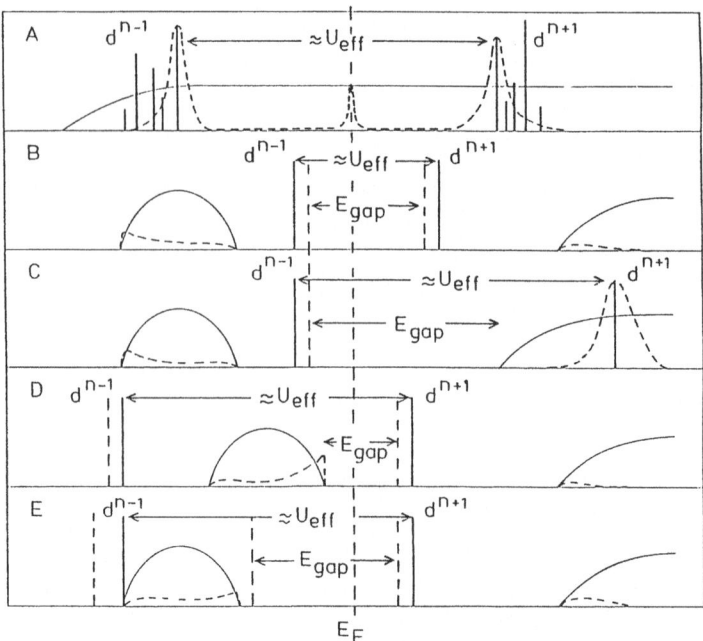

Fig. 1. An artist's concept of the possible situations encountered for
         strongly correlated impurities in solids. Dashed lines indicate
         the effect of hybridization.

different hosts the chemical potential or Fermi level in Fig. 1 can be
shifted. For $U \gg \Gamma$ but $E(d^{n-1})$ or $E(d^{n+1}) \lesssim \Gamma$, we can also have low energy
scale charge fluctuations but only by at most <u>one</u> d or f electron. This
situation leads to the mixed valence and valence fluctuation phenomena
common in many rare earth and transition metal systems. This requires a
study of the more general Anderson Hamiltonian. A theoretical breakthrough
in this problem was accomplished by Gunnarsson and Schönhammer[18] who used
the nature of the high degeneracy of d and f states and expansions in terms
of the inverse of the degeneracy[19] to calculate the density of states resulting from the Anderson model Hamiltonian. This allowed a detailed comparison of theory with both high and low energy scale experimental studies.

It turns out that the Anderson impurity Hamiltonian is also very useful
for describing the electronic structure of pure metals like the rare earth
intermetallics. This then makes direct connection with narrow band phenomena.
The neglect of translational symmetry for the 4f rare earth orbitals appears
to be a valid approximation at least for energy scales above 100 meV. For
the very low energy scale properties, however, the translational symmetry
is certain to become important. This problem of an Anderson or Kondo lattice
is presently under study in several groups and is thought to be important
for describing heavy fermion systems as well as magnetic ordering and superconductivity in transition metal and rare earth compounds.

Customarily the Anderson Hamiltonian has been primarily used for
metallic systems and the Hubbard Hamiltonian for insulators. The reason
for this is partly historical but also because in metallic systems containing
rare earth or transition metals the lowest energy charge fluctuation states
often involve other electrons than the correlated 4f/3d electrons. As mentioned above[8,9] other charge fluctuations than those involving the 3d orbitals
are also important in describing the electronic structure of transition

metal compounds. It has recently been pointed out by Zaanen et al.[20] that the richness of physical properties of 3d transition metal compounds can be explained quite naturally by using an Anderson type Hamiltonian with the conduction band states replaced by the anion valence p orbitals and cation 4s, 4p states, which in the ionic insulators are respectively full and empty. This problem of highly correlated states embedded in a semi-conductor like density of states rather than a metal has been treated before by Haldane and Anderson[21] to describe multiple charged impurity states in semiconductors.

These model Hamiltonian approaches have provided us with a great deal of insight into the electronic structure and physical properties of narrow band systems. They are, however, based on strongly oversimplified descriptions of the electron-electron interaction as well as the one electron dispersion relations and hybridization. For example, the electron-electron interaction is usually replaced by only the on-site interaction with the assumption that all other interactions are incorporated in renormalizations of the parameters in the Hamiltonian. This leads to highly effective parameters which, one might think, could only be obtained from experiment. It is perhaps somewhat surprising that such extreme simplifications can in fact be used to describe a wide range of experimental results. One of the recent successes in this line was indeed to show that both the low and high energy scale properties of some rare earth[18] and transition metal compounds[20] can be explained using the same theory and (nearly) the same parameters.

Theoretically it would be most satisfying to obtain the electronic structure from ab initio calculation. This, however, turns out to be an impossible task and approximations must be made. The development of ab initio methods has, however, made tremendous progress and is now surprisingly accurate in calculations of the ground state properties like the energy, charge density, Fermi surface, crystal structure, lattice parameters, magnetic structure, etc. For narrow band materials, however, ab initio methods have run into problems in describing the excited states like temperature dependencies, bang gaps, optical properties, phase transition etc.

The ab initio methods referred to here are the band theory methods based on the density functional theory and the local density approximation. These methods and their validity to narrow band systems are discussed by several authors in this chapter. The general conclusion reached is that one really should not expect success for the excited states of narrow band materials because the individual wave functions obtained in density functional do not have physical meaning.

A next best scenario could be to try to use ab initio methods to calculate parameters used in model Hamiltonians, which in turn can be used to describe the excited state properties.

The most important development in ab initio methods for solids after the very basic ideas of Bloch[1] and Wilson[2] was the development of density functional theory of Hohenberg, Kohn and Sham[21,22] (see also C.-O. Almbladh in this chapter). These authors showed that the exact ground state energy and charge density can be obtained by solving the one electron Schrödinger equations with an added one particle exchange-correlation potential. The exchange correlation potential can be written as a functional of the electron charge density and a central problem then is to obtain this functional. A much used and extremely successful approximation[23] is the local density approximation (LDA), the limitations of which are discussed by Almbladh in this chapter.

It is generally appreciated that the one particle wave functions and energies in a density functional calculation have no real physical meaning and therefore are not expected to describe the excited states. A question which obtained considerable attention at the meeting was why does (local) density functional theory do so well in describing the Fermi surface of the heavy fermion systems like UPt_3? A question subsequently addressed was: does density functional theory yield the exact Fermi surface? Almbladh argued that the exact description of Friedel oscillation in the screening of a test charge in density functional theory suggests that also the exact Fermi surface is obtained. This, however, leads to a big problem for CoO which in the best of calculations does not exhibit a gap and has a Fermi surface, however, in reality CoO is an insulator. This problem can only be solved if one can somehow obtain a gap in the density functional calculation albeit too small.

Although it has taken a long time before the density functional and model Hamiltonian "clubs" have agreed that both approaches are important, we have finally come to a sufficient mutual appreciation to use the powerful density functional methods to calculate the parameters appearing in model Hamiltonians. How this is done with some examples is described in this chapter by Schönhammer and Gunnarsson as well as McMahan and Martin. In another ab initio method, which is meeting with some success in narrow band materials, one replaces the solid by a cluster and applies configuration interaction approaches as developed in quantum chemistry and molecular chemistry and physics. Especially in narrow band materials in which the translational symmetry of at least some atoms is perhaps of minor importance a replacement of the solid by a cluster seems justifiable. Since one then deals with a finite system with a limited number of electrons, the many electron eigenstates of the cluster can be accurately calculated. Bagus will describe some of these methods below. It turns out that even for strongly localized ionic systems with narrow bands, long range polarization effects are of great importance in describing ionization potentials electron affinities as well as charge transfer energies as shown by Janssen et al.[24] for NiO. Up to this point these long range polarization effects have not been included self-consistently in cluster calculations.

REFERENCES

1. F. Bloch, Z. Phys. 57:545 (1929).
2. A.H. Wilson, Proc. Roy. Soc. A133:458 (1931).
3. H.J. de Boer and E.J.W. Verwey, Proc. Phys. Soc. A49:59 (1937).
4. N.F. Mott, Proc. Phys. Soc. Sect A62:416 (1949).
5. N.F. Mott, "Metal-Insulator Transitions" (Taylor & Francis, London, 1974).
6. J. Hubbard, Proc. Roy. Soc. SerA:277, 237 (1964) and 281:401 (1964).
7. P.W. Anderson, Phys. Rev. 115:2 (1959).
    Solid State Physics 14:99 (1963).
8. A. Fujimori, F. Minami and S. Sagano, Phys. Rev. B29:5225 (1984);
    G.A. Sawatzky and J.W. Allen, Phys. Rev. Lett. 53:2239 (1984).
9. J.H. van Vleck "The Theory of Electric and Magnetic Susceptibilities" Oxford Univ. Press London 1932.
10. J.S. Griffith "The Theory of Transition Metal Ions", Cambridge Univ. Press, London 1961.
11. Y. Tanabe and S. Sugano, J. Phys. Soc. Japan 9:767 (1954);
    S. Sagano, Y. Tanabe and H. Kamimura in "Multiplets of Transition Metal Ions in Crystals", Academic, New York 1970.
12. J.B. Goodenough, Phys. Rev. 100:564 (1955;
    J. Kanamori, J. Phys. Chem. Solids 10:87 (1959).
13. J. Friedel, J. Phys. Radium 19:573 (1958);
    A. Blandin and J. Friedel, J. Phys. Radium 20:160 (1959).
14. P.W. Anderson, Phys. Rev. 124:41 (1961).
15. J.R. Schrieffer and P.A. Wolff, Phys. Rev. 149:491 (1966).

16. J. Kondo, Progr. Theoret. Phys (Kyoto) 28:846 (1962); 32:37 (1964); 34:204 (1965).
17. K.G. Wilson, Rev. Mod. Phys. 47:773 (1975); N. Andrei, K. Furuya and J.H. Lowenstein, ibid 55:331 (1983).
18. O. Gunnarsson and K. Schönhammer, Phys. Rev. Lett. 50:604 (1983); Phys. Rev. B28:4315 (1983).
19. T.V. Ramakrishman in "Valence Fluctuations in Solids" eds. L.M. Falicov, W. Hanke and M.B. Maple, North Holland 1981; P.W. Anderson ibid.
20. J. Zaanen, G.A. Sawatzky, and J.W. Allen, Phys. Rev. Lett. 55:418 (1985). J. Zaanen, and G.A. Sawatzky, Phys. Rev. B (in press).
21. P. Hohenberg and W. Kohn, Phys. Rev. 136:864 (1964).
22. W. Kohn and L.J. Sham, Phys. Rev. A140:1133 (1965).
23. U. van Barth and A.R. Williams in:"Theory of the Inhomogeneous Electron Gas", Eds. S. Lundqvist and M.H. March (Plenum, New York, 1983).
24. G.J.M. Janssen and W.C. Nieuwpoort, Phys. Rev. B (in press).

BANDSTRUCTURE AND LOCALIZED DESCRIPTIONS OF NARROW-BAND SYSTEMS

C.-O. Almbladh

Department of Theoretical Physics, Lund University

Sölvegatan 14A, S-223 62 Lund, Sweden

INTRODUCTION

In this note I would like to give some personal views on the relevance
of bandstructure treatments of narrow-band systems. At first sight one
might think that bandstructure theory is not a very appropriate framework
for quasilocalized electrons. Nevertheless, it has been found that density-
functional (DF) theory in its modern form gives a rather impressive description of ground-state properties for the majority of systems, not only
"in principle" but also in practice via calculations based on the local-
density approximation (LDA). I will here discuss some key concepts in DF
theory with emphasis on physical significance of the eigenvalues which
appear in the DF one-electron-like equations. I will also to some extent
discuss the dynamical selfenergy Σ, which describes the "one-electron"
excitations reached by adding or removing an electron, although there is
very little work done for narrow-band electrons.

The bandstructure schemes are free from adjustable parameters, but
have on the other hand a limited scope. A ground-state calculation can only
be claimed to give properties like spin densities, static polarizabilities,
and Fermi surfaces. The one-electron excitation spectrum obtained from Σ is
closely related to photoemission and inverse photoemission spectra, but for
narrow-band systems it has no simple relations to experiments where more
than one electron or hole are simultaneously created, as for example in
Auger emission. Instead such spectra typically show strong atomic-like
effects like term splittings, resonances, etc., and can often be well described by one-site or impurity models. However, the problem to obtain the
remaining solid-state corrections in a systematic way cannot be regarded
as solved, although a considerable insight has been gained from model
studies. In order to illustrate the advantages and weaknesses of the two
points of view, recent results on transition-metal (TM) oxides will be discussed. I conclude by a few general remarks on the problem why and when
narrow-band electrons behave as localized in different spectroscopies.

GROUND STATE PROPERTIES

I would like to discuss here some key concepts in DF theory,[1,2] and in
particular the relations between excitation energies and the DF one-electron
eigenvalues. There exist several comprehensive reviews[3-9] to which the
reader is referred for more details.

As is well known DF theory gives an in principle exact description of ground-state properties in terms of one-electron-like Schrödinger equations,

$$\left[-\frac{1}{2} \nabla^2 + w(\vec{r}) + V_H(\vec{r}) + v_{xc}(\vec{r}) \right] \phi_i(\vec{r}) = e_i \phi_i(\vec{r}).$$

(1)

The central quantity in this theory is the exchange-correlation energy functional $E_{xc}[n]$, which determines the exchange-correlation potential

$$v_{xc}(\vec{r}) = \delta E_{xc}[n]/\delta n(\vec{r}).$$

(2)

The remaining terms in Eq. (1) represent the electrostatic potential from the nuclei (w) and the electrons (V_H). The ground-state energy E_0 and the electron density n for the interacting N-electron system follow in a simple way from the N lowest self-consistent orbitals in Eq. (1),

$$E_0 = \sum_1^N e_i - \int d^3r \, n \left(\frac{1}{2} V_H + v_{xc} \right) + E_{xc}[n], \quad n(\vec{r}) = \sum_i^N |\phi_i(\vec{r})|^2.$$

DF theory can be extended also to spin-polarized systems.[10,11] In that case the spin-dependent density

$$n_{\alpha\beta}(\vec{r}) = \langle \psi_\alpha(\vec{r})\psi_\beta(\vec{r}) \rangle$$

plays the role of the fundamental variable rather than the particle density. The basic functional E_{xc} now depends on the spin-dependent density, and the one-electron equations are modified accordingly.

In DF theory the correlation problem is hidden in the imprecisely known functional E_{xc}. Clearly the theory would be little more than an elegant mathematical transformation if not, due to fortunate and not fully understood reasons, already the simple local-density approximation usually gives quantitatively accurate results for real solids. The LDA is obtained by ignoring all non-localities in E_{xc}, which then can be expressed in the exchange-correlation energy per particle (ε_{xc}) of a homogeneous system,

$$E_{xc}[n] \simeq \int d^3r \, n(\vec{r})\varepsilon_{xc}(n(\vec{r})).$$

(3)

For properties like electron removal energies and cohesive energies the LDA typically gives errors of the order 1 eV, and thus the approximations cannot compete with more elaborate methods based on configuration-interaction (CI) or many-body perturbations theory in simple systems. Structural and vibrational properties involve only rather slowly varying charge rearrangements, and for these properties the LDA works better. For instance, phonon frequencies in sp-bonded solids can typically be obtained with an accuracy of a few percent with the nuclear charge as the only input.[12]

The central problem of obtaining improvements beyond the LDA has turned out to be a difficult one. Useful insight into why the LDA works so well has been obtained from real-space formulations by Gunnarsson and coworkers,[13] but the most promising new development appears to be the effective gradient scheme due to Langreth, Perdew, and Mehl (LPM).[14,15] This new scheme has been found to give systematic improvements of total energies in atoms, diatomic molecules and solids (for an up-to-date review see Ref. 9). However, there are only minor improvements of densities and eigenvalues for those simple systems where one can compare with accurate results from CI calculations.[16]

Besides the development of better functionals there has also been a renewed interest in foundations of the theory[17] as well as the physical significance of the fictitious eigenvalues appearing in the exact DF one-electron equations. The discussion of eigenvalues was initiated to a large extent by Williams and von Barth[18] and has led to a number of new results. As is well known the eigenvalues as they come out from LDA calculations usually approximate excitation energies in sp-bonded materials reasonably well, whereas problems arise from transition metal and rare-earth systems. Some insight relevant for quasi-localized electrons in narrow bands can probably be obtained from known results for finite systems.[5,18-21] Here one finds that the exact DF eigenvalues lie above the corresponding excitation energies for occupied states and below for unoccupied states and that the differences increase with increasing localization. An exception is the DF eigenvalue for the highest occupied orbital, which agrees with the correct ionization potential (or Fermi energy μ) for all systems, whether finite[19] or extended[2]. The LDA further increases these differences and the highest occupied LDA eigenvalue is far too high. Thus, whereas the LDA gives rather good total energies and densities its description of eigenvalues in localized systems is poor. These inaccuracies may lead to an incorrect population of the orbitals. For instance, negative ions do not bind properly in the LDA or LPM schemes, there are systems where it is impossible to populate the levels according to increasing eigenvalues,[22] and insulators may incorrectly come out as metal in the LDA. Much further work seems to be needed in order to correct for these weaknesses.

To answer the principal question whether or not the exact DF eigenvalues give the correct shape of Fermi surfaces in metals we consider an impurity in an otherwise perfect host. The experimental and DF Fermi surfaces of the host are given by

$$\varepsilon_k = \mu \qquad\qquad (4)$$

and

$$e_k = \mu \qquad\qquad (5)$$

in terms of quasiparticle energies (ε_k) and DF eigenvalues (e_k). From the Luttinger theorem[23] we know that both surfaces enclose the same volume in k space, and they have the same symmetry. The density change induced by the impurity exhibits Friedel oscillations at far distances, and these oscillations can be shown to give a direct measure of the dimensions of the quasiparticle Fermi surface in Eq. (4). Because DF theory is postulated to reproduce the exact density these oscillations are also a consequence of the DF Fermi surface, leading to the conclusion that the two surfaces are identical.[24,25]

In real metals the LDA Fermi surfaces usually are quite reasonable although not strikingly accurate. Thus, in simple metals the LDA gives small but significant deviations, corresponding to eigenvalue differences of the order a few mRy.[26] For Cu MacDonald et al.[27] obtained a Fermi surface of comparable accuracy in terms of eigenvalue shifts, whereas the LDA generally gives larger error for the transition metals. However, a rather remarkable agreement has been obtained for the heavy fermion system UPt_3, where the LDA gives the correct Fermi surface within 1 mRy.[28]

The fundamental bandgap in insulators was long argued to be given by the corresponding gap

$$\Delta e_{cv} = e_c - e_v \tag{6}$$

in the exact DF eigenvalue spectrum. The reason for this belief was that the true gap involves only ground-state energy differences obtained by adding and subtracting one electron. It was independently proposed by Perdew and Levy[29] and by Sham and Schlüter[30] that there in general is a finite difference between the DF gap and the true excitation gap ($\Delta \varepsilon_{cv}$),

$$\Delta \varepsilon_{cv} = \Delta e_{cv} + \Delta, \tag{7}$$

and from exact solutions of models[24],[31] we know that their conjecture is indeed correct. I would like to briefly review the models before commenting on recent numerical results on real semiconductors. Almbladh and von Barth[24] consider a model insulator with a very large lattice spacing and obtain an exact solution in terms of the known results of the isolated atoms, whereas Gunnarsson and Schönhammer[31] consider a Hubbard chain, solve the model numerically and extrapolate to infinitely many sites. These solutions, however, do not indicate any simple mechanism for the band gap correction Δ. Thus, for a sparse solid of Be atoms there is a large gap correction Δ (5.8 eV) and the LDA and exact DF bandgaps almost coincide, whereas for a sparse Ne solid almost half of the bandgap discrepancy (9.7 eV) is removed when going from the LDA to exact DF theory. The gap in the Be system involves only excitations within the same shell (2s, 2p) and has some qualitative similarity with the 3d gap in magnetic insulators discussed in the next section. On the other hand the Hubbard model gives only small gap corrections and attributes most of the discrepancy to the LDA. Much further work is needed in order to understand the DF description of band gaps. The true gap, being a ground state property, would in principle be obtainable from DF theory, and a particularly important problem is to develop practical schemes for this.

The correct one-electron excitations follow from a one-electron potential, the dynamical selfenergy $\Sigma(\vec{r}, \vec{r}', \varepsilon)$. To obtain these energies one solves the Dyson equation,

$$\left[-\frac{1}{2} \nabla^2 + w + V_H + \Sigma(\varepsilon_i) \right] \chi_i = \varepsilon_i \chi_i, \tag{8}$$

and Hybertsen and Louie[32] and Godby et al.[33] have recently managed to obtain realistic solutions for several semiconductors and insulators. They use the simplest non-trivial approximations to Σ, the "GW" approximation[34] and obtain results which agree very well with the measured bands. These nice results suggest that the same approximation could be useful also for d-electron systems.

Godby et al. also shed light on the DF eigenvalue spectrum by constructing the ground-state potential v_{xc} which reproduces the GW density profile. The corresponding DF eigenvalues are found to be rather close to their LDA counterparts, and in particular most of the difference between the LDA gap and the excitation gap (80% in Si) attributed to the gap correction Δ. The reliability of these results clearly relies on the accuracy of the GW densities. In order to obtain these densities the full Green's function must be constructed for all energies, and provided Godby et al. have managed to do this in an accurate way there is a number of independent arguments which support their claim to have obtained significant improvements. In atoms the GW scheme treats the exchange exactly, and correlation is known to have only a minor influence on the correct DF eigenvalues.[16,21] Consequently the GW density-functional scheme would remove the major part of the LDA eigenvalue errors here. In the case of metals it is not difficult to see that an approximate v_{xc} gives the same Fermi surface as the selfenergy from which it was derived, and as the GW scheme is known to yield rather accurate Fermi surfaces[26] we again see that the DF eigenvalues have been improved. We next turn to high-density systems. Long ago Ma and Brueckner[35] identified those terms which give the exact static charge response function in the high-density limit, and these terms may actually be considered arising from the selfconsistent GW scheme. Thus, this approximation gives the correct static polarizability for all q at high densities. We finally note that Godby and coworkers' approximation to v_{xc} has the correct asymptotic behaviour far outside solid surfaces and far away from finite systems.[19,36]

BAND AND LOCALIZED APPROACHES TO TRANSITION-METAL COMPOUNDS

The 3d transition-metal monoxides MnO, FeO, CoO and NiO, which are all antiferromagnetic insulators, are interesting and rather extreme systems for testing band theory. They all involve partially filled d bands, which in the most naive band picture would lead to metallic behaviour. It has long been realized that in these materials the Coulomb correlation tends to outweigh the band dispersional effects, and these ideas form the basis for their description as "Mott-Hubbard" insulators.[37] The smallness of the d-band widths also make localized impurity models rather natural, at least for properties which do not involve cooperative phenomena and charge transport.

Accurate LDA calculations have been performed for the entire series of monoxides[38-40] and give in fact a rather useful description at least for ground-state properties. Thus, the LDA predicts the correct spin structure, even the magnetization direction is correct,[38] it accounts well for local moments, cohesive energies, and lattice parameters. In particular the abrupt lattice change when going from the non-magnetic VO to MnO is well described.[39,40] Also Néel temperatures come out rather well from KKR-CPA calculations.[41] In short, all properties directly related to the spin densities and total energy of the N-electron system are well described.

The predictions of band gaps, however, are not as good. Although NiO and MnO come out as insulators the LDA fails to give a gap in FeO and CoO. As discussed in the previous section the local-density approximation to the DF eigenvalues is probably not very accurate for narrow-band electrons, and as far as Fe and Co are concerned there is an eigenvalue-related failure also in the LDA description of the free atoms.[22] The best studied material is NiO, and in this case the LDA gap (0.6 eV) is much smaller than the gap measured by angle-resolved photoemission[42] and Bremsstrahlung[43] experiments (4.3 eV). Of course the calculated gap approximates the (unknown) DF gap, not the experimental one, and most results available thus far suggest an important gap correction. However, DF theory can be stretched to give estimates also of certain excitation energies, and by adding a selfenergy correction estimated by ΔSCF calculations to the d-electron energies Kübler and Williams[38] obtain a reasonably good agreement with the measured density of states.[42,43] Similar ideas have also been applied by Norman et al.[44]

In order to describe the excitations in a more systematic way from the bandstructure point of view, methods based on a selfenergy would be required. There is very little experience for d-electron systems, but we saw in the previous section that already the GW approximation seems to give rather accurate quasiparticle energies in sp-bonded materials both for band states[32,33] and for states localized at surfaces.[45] Even for completely localized core electrons the major part of the relaxation energy is obtained via the Dyson-equation eigenvalue.[46] It would not be quite unreasonable to expect that the same approximation could be useful also for quasiparticle ("relaxed") bands in d-electron systems, in particular since the major part of the corrections to the LDA gap can be accessed by simple ΔSCF calculations. The GW scheme is mainly an approximation for the quasiparticles, where it at least in some limiting cases has a variational character, but in order to describe satellite parts one most probably needs more refined approximations. An interesting approach has been used by Liebsch[47] in the case of metallic Ni, Liebsch models the Ni metal by a Hubbard Hamiltonian and keeps the full dependence of the Coulomb intra-atomic interaction U on the d-level magnetic sublevels. This leads to a selfenergy Σ whose properties are set both by atomic features through U and band features through the unperturbed one- and two-electron propagators. These ideas can probably be pushed further by utilizing results from both localized models and self-consistent bandstructure calculations in the construction of a better self-energy.

Fujimori et al.[48] treat the problem from exactly the opposite viewpoint. They ignore completely all band and multisite aspects and model the systems by small clusters. This is of course an enormous simplification, and the system can now be handled by configuration-interaction methods with parameter partly taken from experiment. The main ingredients in their theory are the d-d Coulomb integrals, d - ligand-2p hybridization, and crystal-field splitting. The picture that emerges from their studies is that the main d-bands involve large p-to-d screening from neighboring ligand 2p orbitals, whereas the shake-up satellites involve mainly unscreened d electrons or holes. For NiO thus the main "band" in the localized picture has a large oscillator strength on $d^8\underline{L}$ configurations ($\underline{L}$ represents a ligand hole), whereas the satellites involve mainly d^7 configurations. This interpretation is supported by resonant photoemission data, which in the simplest picture would involve a competing channel $d^8 \rightarrow 3\underline{p}d^{10} \rightarrow d^7\epsilon\ell$ opening up at the 3p threshold. One would thus expect the satellite to be resonantly enhanced at the 3p threshold, and this is in accord with both experiments[49] and more detailed calculations.[48]

The finite-cluster description of TM compounds has been refined by Zaanen, Sawatzky, and coworkers[50] who account for the finite bandwidths of the ligand 2p states ($\sim$ 3.8 eV in Ni). Their treatment leads to Anderson impurity models and efficient methods for obtaining the excitations have been developed by Gunnarsson and Schönhammer.[51] In this way Zaanen et al. can account for the insulating and metallic behaviour of a large class of TM compounds and also their high-energy spectra[52] in terms of a few phenomenological parameters. The works by Fujimori, Zaanen and coworkers lead to modifications of the original Mott-Hubbard picture of the TM compounds. Thus, when the Coulomb U is large compared to the charge-transfer energy (Δ) as in NiO, the d-states are straddeling the ligand p levels, and the gap is essentially determined by Δ rather than U.

The CI treatment of cluster or impurity models gives directly oscillator strengths and thus a very explicit picture of the spectroscopic properties. As far as the one-electron excitations are concerned the selfenergy of course contains similar information, but away from the band edges a knowledge of just the quasiparticle solutions to the Dyson equation does not suffice for determining e.g. charge transfers. Are then the pictures emerging from the band and cluster models fundamentally inconsistent? Assuming that Kübler and Williams'[38] selfenergy-corrected bands are close to the correct ones the answer appears to be no. To obtain the correct charge-transfer character of the gap from bandstructure theory, quasiparticle orbitals would be needed rathern than their LDA counterparts, but the large p-to-d screening found in the cluster models is brought out also in simple ΔSCF calculations.[44] A band description of the quasiparticle excitations give results which are difficult to obtain by other methods. For instance, the position and width of the ligand p bands come out well but are input rather than output in the localized models.

The optical properties of NiO, and in particular the origin of the fundamental absorption edge at ~4 eV, is an old and controversial subject.[53] The proponents of localized models[43,48] suggest that the edge arises from interatomic d-d transitions mediated by the intervening ligand orbitals, and Kübler and Williams[38] suggest that the origin could be transitions from 2p ligand states to low-lying d states screened also by the lattice, followed by auto-ionization to free carrier states. Neither of the suggested mechanisms are supported by any estimates of matrix elements. It would appear that transitions from highly correlated states into unbound free carrier states are difficult to account for in both cluster models and LDA band-structure schemes. Hopefully calculations on larger clusters or on correlated bandstructure could shed further light on this old problem.

CONCLUDING REMARKS

The example of TM compounds reveals both the strengths and weaknesses of DF theory in its local-density form. The total energies and spin densities are given quite accurately whereas there are problems with the eigenvalues. These inaccuracies call for better approximations to the DF ground-state potential, and Sham and Schlüter's idea to express it in the dynamical selfenergy could be a useful guide for developing better schemes. In these compounds the LDA eigenvalues evidently are rather poor approximations to d-electron (frozen-lattice) excitation energies, but how much of the discrepancy is due to the LDA is difficult to say at this stage. On the other hand the selfenergy corrections to the LDA seems to be well accounted for by simple ΔSCF calculations.

There appears to be no principal conflict between the localized and band descriptions of the quasiparticle excitations in the cases discussed here, but in order to really bridge the gap between band and localized models for narrow-band systems one needs a thorough understanding of how localized behaviour occurs based on non-perturbative solutions. Early works by Cini[54] and Sawatzky[55] explain nicely the strong quasi-atomic effects in Auger spectra. Their analysis applies to the case of completely filled d-bands, where in a Hubbard-model description the full many-electron nature of the problem can be circumvented. However, in order to account for partially filled bands, core-hole effects and hybridization, one is led to truly interacting models. Important progress has been made primarily for Anderson impurity models via expansions in the reciprocal degeneracy and basis-set techniques.[51] A periodic model, which is more relevant for the discussion here, is inevitably much harder to treat, but some insight has been gained by direct diagonalization of small chains with typically 4-8 sites.[56] The number of sites is really too small to allow for extrapolations to the N=∞ limit, but the exact results could give guidance for developing new approximations.

REFERENCES

1. P. Hohenberg and W. Kohn, Phys. Rev. 136:B864 (1964).
2. W. Kohn and L.J. Sham, Phys. Rev. 140:A1133 (1965).
3. A.K. Rajogopal, in "Advances in Chemical Physics", edited by I. Prigogine and S.A. Rice, 41:59, Wiley, New York (1980).
4. "Theory of the Inhomogeneous Electron Gas", edited by S. Lundqvist and

N.H. March, Physics of Solids and Liquids Series, Plenum, New York (1983).

5. U. von Barth, Density Functional Theory for Solids, in "The Electronic Structure of Complex Systems", edited by P. Phariseau and W.M. Temmerman, NATO ASI Series B: Physics, Vol. 113, p.67, Plenum, New York (1984).

6. "Density Functional Methods in Physics" edited by R.M. Dreizler and J. da Providencia, NATO ASI Series B: Physics, Vol. 123, Plenum, New York (1985).

7. J. Callaway and N.H. March, Density Functional Methods: Theory and Applications, in "Solid State Physics", edited by H. Ehrenreich, F. Seitz, and D. Turnbull, Vol. 40, Academic Press, New York (1984).

8. O. Gunnarsson and R.O. Jones, review article (1986).

9. U. von Barth, Chem. Scripta 26:449 (1986).

10. U. von Barth and L. Hedin, J. Phys. C 5:1629 (1972).

11. A.K. Rajagopal and J. Callaway, Phys. Rev. B 7:1912 (1973).

12. See e.g. A.K. McMahan, M.T. Yin, and M.L. Cohen, Phys. Rev. B 24:7210 (1981).

13. O. Gunnarsson, M. Jonson, and B.I. Lundqvist, Phys. Rev. B 20:3136 (1979); O. Gunnarsson and R.O. Jones, Physica Scripta 21:394 (1980).

14. D.C. Langreth and J.P. Perdew, Phys. Rev. B 15:2884 (1977); D.C. Langreth and J.P. Perdew, Phys. Rev. B 21:5469 (1980).

15. D.C. Langreth and M.J. Mehl, Phys. Rev. B 28:1809 (1983).

16. A.C. Pedroza, Phys. Rev. A 33:804 (1986); C.-O. Almbladh and S. Svendsen (unpublished).

17. M. Levy, Proc. Natl. Acad. Sci. USA, 76:6062 (1979); E.H. Lieb, Int. J. Quantum Chem. 24:243 (1983).

18. A.R. Williams and U. von Barth, Applications of Density-Functional Theory to Atoms, Molecules and Solids, in Ref. 4 p.189.

19. C.-O. Almbladh and U. von Barth, Phys. Rev. B 31:3231 (1985).

20. U. von Barth and R. Car, unpublished work on exact DF potentials.

21. C.-O. Almbladh and A.C. Pedroza, Phys. Rev. A 29:2322 (1984).

22. J.F. Janak and A.R. Williams, Phys. Rev. B 23:6301 (1981).

23. J.M. Luttinger, Phys. Rev. 119:1153 (1960).

24. C.-O. Almbladh and U. von Barth, Density functional theory and excitations energies, in Ref. 6, p.209.

25. C.-O. Almbladh and U. von Barth, to be published.

26. A.H. MacDonald, J. Phys. F 10:1737 (1080).

27. A.H. MacDonald, J.M. Daams, S.H. Vosko, and D.D. Koelling, Phys. Rev. B 25:713 (1982).

28. G.S. Wang, M.R. Norman, R.C. Albers, A.M. Boring, W.E. Picket, H. Krakauer, and N.E. Christensen, Phys. Rev. B, to be published.

29. J.P. Perdew and M. Levy, Phys. Rev. Lett. 51:1844 (1983).

30. L.J. Sham and M. Schlüter, Phys. Rev. Lett. 51:1888 (1983).

31. O. Gunnarsson and K. Schönhammer, Phys. Rev. Lett. 56:1968 (1986); K. Schönhammer and O. Gunnarsson, to be published.

32. M.S. Hybertsen and S.G. Louie, Phys. Rev. B 43:5390 (1986)

33. R.W. Godby, M. Schlüter, and L.J. Sham, Phys. Rev. Lett. 56:2415 (1986); R.W. Godby, M. Schlüter, and L.j. Sham, Phys. Rev. B 35:4170 (1987); R.W. Godby, M. Schlüter, and L.J. Sham, to be published.

34. L. Hedin, Phys. Rev. 139:A796 (1965).

35. S.-K. Ma and K. Brueckner, Phys. Rev. 165:18 (1968).

36. L.J. Sham, Phys. Rev. B 32:3876 (1985).

37. N.F. Mott, Proc. Phys. Soc. A 62:416 (1949); J. Hubbard, Proc. Roy. Soc. A 277:237 (1964); ibid 281:401 (1964).

38. K. Terakura, T. Oguchi, A.R. Williams, and J. Kübler, Phys. Rev. B 30:4734 (1984); K. terakura, A.R. Williams, T. Oguchi, and J. Kübler, Phys. Rev. Lett. 52:1830 (1984); J. Kübler and A.R. Williams, J. Magn. Magn. Mater. 54-57:603 (1986).

39. J. Yamashita and S. Asano, J. Phys. Soc. Jpn. 52:2514 (1983).

40. O.K. Andersen, H.L. Skriver, H. Nohl, and B. Johansson, Pure & Appl. Chem. 52:93 (1979).
41. T. Oguchi, K. terakura, and A.R. Williams, Phys. Rev. b 28:6443 (1983).
42. J.M. McKay and V.E. Henrich, Phys. Rev. Lett. 53:2243 (1984).
43. G.A. Sawatzky and J.W. Allen, Phys. Rev.Lett. 53:2339 (1984).
44. M.R. Norman and A.J. Freeman, Phys. Rev. B 33:8896 (1986).
45. M.S. Hybertsen and S. Louie, to be published.
46. L. Hedin, Arkiv Fysik 30:231 (1965).
47. A. Liebsch, Phys. Rev. B 23:5203 (1981).
48. A. Fujimori and F. Minami, Phys. Rev. B 30:957 (1984).
49. M.R. Thuler, R.L. Bendow, and Z. Hurych, Phys. Rev. B 27:2082 (1982); S.J. Oh, J.W. Allen, I. Lindau, and J.C. Mikkelsen, Phys. Rev. B 26:4845 (1982).
50. J. Zaanen, G.A. Sawatzky, and J.W. Allen, Phys. Rev. Lett. 55:418 (1985) and J. Magn. Magn. Mat. 54-57:607 (1986).
51. O. Gunnarsson and K. Schönhammer, Phys. Rev. Lett. 50:604 (1983); Phys. Rev. B 28:4315 (1983); ibid 31:4815 (1985).
52. G. van der Laan, J. Zaanen, G.A. Sawatzky, R. Karnatak, and J.M. Esteva, Phys. Rev. B 33:4253 (1986).
53. For a review see B.H. Brandow, Adv. Phys. 26:651 (1977).
54. M. Cini, Solid State Commun. 20:605 (1976).
55. G.A. Sawatzky, Phys. Rev. Lett. 39:504 (1977).
56. R. Julien and R.M. Martin, Phys. Rev. B 26:6173 (1981); A.M. Olés, G. Tréglia, D. Spanjaard, and R. Julien, Phys. Rev. B 34:5101 (1986); R.H. Victoria and L.M. Falicov, Phys. Rev. Lett. 55:1140 (1985); A. Reich and L.M. Falicov, Phys. Rev. B 34:6752 (1986).

THE MOLECULAR ORBITAL CLUSTER MODEL APPROACH
TO ELECTRONIC STRUCTURE

P. S. Bagus

IBM Research
Almaden Research Center
650 Harry Road
San Jose, California 95120-6099

The cluster model approach involves choosing a finite, and not exceptionally large, number of atoms to represent an extended, condensed matter, system of interest. Typically, the number of atoms included in the cluster is $\sim$20, see for example Refs. 1 and 2, but larger numbers have been used.[3] The electronic wave function for this cluster is obtained using the methods of ab initio molecular orbital, MO, theory; methods which were developed for Quantum Chemistry.[1,4] When electron correlation effects are not included, a self-consistent field, SCF, wave function is used; when correlation effects are included, either multi-configuration SCF, MCSCF, or general configuration interaction, CI, wave functions are used. One particularly attractive feature of this approach is that approximations are included in a controlled fashion. Parameters which are adjusted to make the computed results fit experimental data are not used. In principle, the exact cluster wavefunction may be obtained by using a sufficiently large CI. The approach is truly ab initio.

The principal virtue of the application of the MO cluster model to extended systems is that it can be used to make detailed studies of localized properties. For this reason, it has been used extensively in two general areas: (1) chemisorption phenomena[5] and (2) properties of ionic crystals, a very few examples of studies related to ionic crystals are given in Refs. 2.6-8 The fact that ionized core levels can naturally be localized in an MO approach[9] is an obvious example of how the approach is suited to treatment of localized properties. The focus of the MO cluster model on systems where localization is important make it valuable for the study of narrow band phenomena. Some of the most important features of the MO cluster model are briefly summarized in the following.

With Quantum Chemistry methods, it is possible to identify and treat individually particular electron correlation effects; the importance of a specific correlation effect can be tested and evaluated. For the metal-carbonyl bond, for example, a major correlation effect is to increase the metal $d\pi$ to CO $2\pi^*$ covalent dative bonding or back-donation.[10] While this correlation affects the precise values of the bond energy and distance, it does not change the general form of the bond from the SCF description.

Another important example of a specific correlation effect is for the shallow core level ionization of transition metal atoms.[6,7,11] The specific case that is summarized here is for 3s ionization of the Mn^{2+} ion in MnF_2 or MnO ionic crystals. The experimental observation[11] is that there is a doublet for the 3s photoemission peak. The isolated Mn^{2+} ion, which is the simplest model for Mn in an ionic crystal, has a $3d^5$ occupation with the spin of all five d electrons aligned parallel; i.e., $d\alpha^5$ with ^6S local multiplicity. The first way to think of the origin of the 3s doublet is that the unrestricted Hartree-Fock,

UHF, ionization energy for a $3s\beta$ electron is lower than that for a $3s\alpha$ electron by five times the 3s-3d exchange integral. An improvement over the UHF view takes into account that in the final ionic state the open 3s and 3d shells form many-electron multiplets, ^7S and ^5S; in this case, properly coupled restricted Hartree-Fock, RHF, wave functions are necessary. The RHF multiplet theory prediction[6] for the splitting of the 3s doublet in the free Mn^{2+} ion is 14 eV, more than a factor of 2 larger than the splitting observed. The relative intensity of the two peaks predicted by the RHF theory is 1.4:1, ^7S more intense than ^5S, while the observed intensity ration is 2:1. The UHF theory has similar problems. One possible conclusion is that an additional major physical effect is present for the MnF_2 and MnO crystals and not for the free Mn^{2+} ion. This is not the case; in fact, a key correlation effect has been neglected in the RHF and UHF treatments. This effect arises from the near degeneracy of the 3s, 3p, and 3d levels; for an H atom these levels are exactly degenerate. This near degeneracy is taken into account with an MCSCF wave function that permits all possible distributions of the 12 M shell electrons of the 3s ionized Mn^{2+} into the 3s, 3p, and 3d shells. For this MCSCF wave function, the predicted[7] splitting and relative intensity are now in good agreement with experiment. The MCSCF multiplet theory also predicts low lying satellites of the ^5S state. These satellites are observed providing further confirmation of the multiplet splitting origin of the 3s doublet and associated satellites. This origin means that 3s photoionization is a source of spin-polarized electrons[12] internal to the material being ionized. Spin-polarized photo-electron diffraction,[12] using this internal electron source, is a powerful and unique probe of local magnetic order in an anti-ferromagnet.

In both of these examples, correlation is important. Note that MO cluster model theory does not require the use of correlation as a magic black box that provides the correct answer. It is possible, as has been shown, to use the theoretical analysis to relate the correlation to physical behavior and mechanisms and to understand the limitations of SCF theory.

Another advantage of the MO cluster model is that the electronic wave functions can be analyzed to clearly reveal the nature of the interaction and chemical bonding between component units of the cluster. Particular questions of concern are the importance of various charge redistributions which occur when a chemical bond is formed and the characterization of a bond as being ionic or covalent. Two newly developed ways of getting definitive answers to these questions are, all too briefly, summarized below.

In order to evaluate the contribution of different types of charge rearrangements to the binding energy and other properties of an interaction, the constrained space orbital variation, CSOV, method[13] has been developed. The CSOV method makes it possible to separate the various kinds of intra-unit and inter-unit charge redistributions involved in a chemical interaction and to determine the importance of each. The CSOV has been used to examine the nature of metal-CO bonding. For the inter-unit charge rearrangements, it shows that the metal π back-donation to $CO(2\pi^*)$ is substantially more important than the CO σ donation to the metal.[13,14] For CO chemisorbed on a metal surface,[1,14] the metal charge has a significant intra-unit polarization to reduce the overlap repulsion with CO. The CSOV analysis has also been applied to understand the vibrations[15] and photoemission[16] of chemisorbed CO. The features of the interaction described above lead to a surprising re-interpretation of the significance of these phenomena. The generalization of the CSOV analysis to other system is straightforward.[17]

The projection of the orbitals of a component onto the wave function of the total cluster provides an indication of the importance of these orbitals in the full cluster. The projection operator, P, for an orbital ϕ_i is $P = \phi_i\phi_i^\dagger$; the cluster expectation value of P is a measure of the extent to which ϕ_i is occupied in the cluster. Projection has shown[18,19] that CN is ionic, CN^-, when it interacts with a metal atom or surface; on the other hand, the CN interaction with an organic unit is covalent. In another application, projection, combined with analysis of the dipole moment, shows that O chemisorbed on Ni(100) is strongly negatively charged; an estimate of its ionicity is -1.5. Projection can be used to determine the 3d shell occupation of a transition metal atom. In particular for ionic crystals, it can be used to identify the difference in the d occupation for atoms which have different nominal ionicities; e.g., $+2$ and $+3$.

In summary, it is important to stress that MO clusters can provide detailed electronic structure information about the localized properties of extended systems. The object of cluster studies is not to exactly reproduce experimental results; the convergence of cluster properties with respect to size[1,2] makes this difficult. The object is to identify and to characterize the physical mechanisms which lead to the observed results. For several reasons, the ab initio MO cluster method is very well suited for this. (1) It does not use parameters adjusted to experiment. (2) Correlation effects can be included in a controlled fashion. And (3), new techniques for the analysis of the electronic wave functions make it possible to quantify the importance of various conceptual models used to interpret experimental data.

REFERENCES

1. K. Hermann, P. S. Bagus, and C. J. Nelin Phys. Rev. B (in press).
2. J. Q. Broughton and P. S. Bagus, Phys. Rev. B 30:4761 (1984); ibid. (in press).
3. See, e.g., J. L. Whitten, Phys. Rev. B 24:1810 (1981); J. L. Whitten and T. A. Pakkanen, Phys. Rev. B 21:4357 (1980); and C. R. Fischer and J. L. Whitten, Phys. Rev. B 30:6821 (1984).
4. H. F. Schaefer III, "The Electronic Structure of Atoms and Molecules," Addison-Wesley, Reading (1972).
5. P. S. Bagus, C. J. Nelin, and K. Hermann, Austral. J. Phys. 39:731 (1986).
6. A. J. Freeman, P. S. Bagus, and J. V. Mallow, Int. J. Magnetism 4, 35 (1973).
7. P. S. Bagus, A. J. Freeman and F. Sasaki, Phys. Rev. Lett. 30:850 (1973).
8. G. J. M. Janssen and W. C. Nienwpoort, Phil. Mag. B51:127 (1985); Solid State Ionics 16:29 (1985).
9. P. S. Bagus and H. F. Schaefer III, J. Chem. Phys. 56:224 (1972).
10. P. S. Bagus, C. J. Nelin and C. W. Bauschlicher, J. Vac. Sci. Tech. A 2:905 (1984); C. W. Bauschlicher, P. S. Bagus, C. J. Nelin, and B. O. Roos, J. Chem. Phys. 85:354 (1986).
11. S. P. Kowalczyk, L. Ley, R. A. Pollack, F. R. McFeely, and D. A. Shirley, Phys. Rev. B 7:4009 (1973); C. S. Fadley and D. A. Shirley, Phys. Rev. A 2:1109 (1970).
12. B. Sinkovic, B. Hermsmeir, and C. S. Fadley, Phys. Rev. Lett. 55:1227 (1985); B. Sinkovic and C. S. Fadley, Phys. Rev. B 31:4665 (1985).
13. P. S. Bagus, K. Hermann, and C. W. Bauschlicher, J. Chem. Phys. 80:4378 (1984); ibid. 81:1966 (1984).
14. C. W. Bauschlicher and P. S. Bagus, J. Chem. Phys. 81:5889 (1984).
15. P. S. Bagus and W. Müller, Chem. Phys. Lett. 115:540 (1985); W. Müller and P. S. Bagus, J. Vac. Sci. Tech. A 3:1623 (1985).
16. P. S. Bagus and K. Hermann, Phys. Rev. B 33:2987 (1986).
17. L. G. M. Pettersson and P. S. Bagus, Phys. Rev. Lett. 56:500 (1986).
18. P. S. Bagus, C. J. Nelin, W. Müller, M. R. Philpott, and H. Seki, Phys. Rev. Lett. 58:559 (1987).
19. C. J. Nelin, P. S. Bagus and M. R. Philpott, J. Chem. Phys. (in press).

VARIATIONAL MONTE-CARLO METHOD FOR STRONGLY INTERACTING ELECTRONS

T.M. Rice

Theoretische Physik

ETH-Hönggerberg, CH-8093 Zurich, Switzerland

Although the variational Monte- Carlo method has been successfully used for a number of manybody problems such as normal state of ^{3}He its application to the problem of electrons in solids has been rather limited and it is only quite recently that it has been tried on the Hubbard and periodic Anderson Hamiltonians. Its advantages are clear. It allows the essentially exact evaluation of matrix elements of wavefunctions which have strong shortrange correlations explicitly built in to them. This then enables various approximate schemes to be tested and the development of intuition about the properties of these wavefunctions.

The success of the method of course depends on the choice of wavefunction. So far only wavefunctions describing paramagnetic Fermi liquid or insulating states have been explored and the interest has focussed on the localized or almost localized regimes. The former describes a Mott insulator where the energy gap is due to the correlations while the latter describes a Fermi liquid in which the strong local correlations modify the properties of the Fermi liquid. Examples of almost localized Fermi liquids are the metallic state of V_2O_3 under pressure near to its Mott transition, the normal state of ^{3}He near to the solidification transition and the heavy electron metals. In the first two examples the system is close to localization and can be driven through the transition by changing the pressure. In the heavy electron metals it is the small number of electrons promoted out of the f-band which are responsible for the Fermi liquid character and the small parameter which controls the large mass enhancement.

The combination of strong short range correlations and a Fermi surface which satisfies Luttinger's theorem, which requires that the volume enclosed by the Fermi surface does not change with interaction strength, is most simply expressed in the Gutzwiller form of wavefunction,

$$|\Psi_G> = P(d)|\Psi_{Band}>$$

where $|\Psi_{Band}>$ is a one-electron band wavefunction made up from a single band (Hubbard model) or more than one hybridized bands (Anderson model). $P(d)$ is a projection operator which enforces the strong local correlation by restricting the charge fluctuations so that the average concentration of doubly occupied sites has the value, d. It is the combination of order in $\vec{k}$-space expressed through the Fermi surface and coherent Fermi liquid properties and the strong local correlation in $\vec{r}$-space which makes this problem difficult. If we extend $|\Psi_G\rangle$ out explicitly in a $\vec{r}$-space representation then only a restricted number of all the real space distributions implicit in $|\Psi_{Band}\rangle$ are allowed, each weighted by a product of Slater determinants.

Horsch and Kaplan[1] were the first to recognize that the averaging process over these Slater determinants could be carried out numerically using a Monte-Carlo procedure. These calculations require repeated operations with Slater determinants and the inverse update technique introduced by Ceperley et al.[2] is an efficient numerical technique to handle these. In certain cases, Gross et al.[3] have made use of the fact that the Slater determinant has a Vander Monde form and so can be represented as a single product which allows even faster numerical evaluations. The first calculations by Horsch and Kaplan[1] were for the case of a 1/2-filled single band Hubbard model in one and two dimensions with system sizes up to $\approx 10^2$ sites. Shiba[4] has studied the Anderson model in one dimension. More detailed studies of the Hubbard model have been made by Gros, Joynt and Rice[3], by Yokoyama and Shiba[5] and by Einarsson[6].

The method is suited to calculate the ground state and various properties of the ground state. Thus it can be used to obtain the ground state energy and its dependence on the filling factor and the concentration of doubly occupied states. As such it can be used to test approximate schemes such as those originally proposed by Gutzwiller[7] or the various slave boson approximation schemes[8] for the expectation value of the kinetic energy, etc. Other properties that are easily accessible are the equal time spin-spin correlation function[1,3-6], which has a strong dependence[9] on the choice of the band function $|\Psi_{Band}\rangle$, the uniform spin susceptibility[3] and the quasiparticle effective mass[3]. At present the method has been restricted to these ground state properties and so far quantities such as $\vec{q}$-dependent magnetic susceptibility and the description of higher temperature thermodynamics have proved beyond the capabilities of the method.

Although the information that we can obtain is rather restricted, the method is one of the few which allows us to treat the simultaneous presence of coherence in $\vec{k}$-space and strong local correlations in $\vec{r}$-space. As such it is complementary to the direct Monte-Carlo simulations of Hirsch and coworkers[10] which are best at high temperatures.

REFERENCES

1. P. Horsch and T.A. Kaplan, J. Phys. C16 L1203 (1983).
    ibid, Bull Am. Phys. Soc. 30:513 (1985).
2. D. Ceperley, G.V. Chester, and M.H. Kalos, Phys. Rev. B16:3081 (1977).
3. C. Gros, R. Joynt, and T.M. Rice, Phys. Rev. B (in press) .
4. H. Shiba, J. Phys. Soc. Japan 55:2765 (1986).
5. H. Yokoyama and H. Shiba (preprint).
6. T. Einarsson, Diplomarbeit ETH-Zurich (1987); and to be published.
7. M.C. Gutzwiller, Phys. Rev. Lett. 10:159 (1963);
    ibid Phys. Rev. A134:923 (1964); ibid Phys. Rev. A137:1726 (1965).
8. G. Kotliar and A.E. Rickenstein, Phys. Rev. Lett. 57:1362 (1986).
9. T.M. Rice, C. Gros, R. Joynt, and M. Sigrist, Proc. 5th Int. Conf. on Valence Fluctuations, Bangalore 1987 (to be published).
10. J.E. Hirsch, Phys. Rev. Lett. 54:1317 (1985) and references therein.

THEORY OF MOTT INSULATORS

Baird Brandow

Theoretical Division
Los Alamos National Laboratory
Los Alamos, NM 87545 USA

We review a conceptual picture of Mott insulators, based on strong orbital as well as spin polarization in unrestricted Hartree-Fock theory, with screened intra-atomic parameters. This picture is used to interpret recent works. The mechanism for the breakdown of local-density band theory is carefully discussed.

INTRODUCTION

The subject of Mott insulators has always been notorious for controversy. The depth of confusion can be sensed from the fact that, when I reviewed this subject a decade ago (Ref. 1, hereafter called I), there was still no general concensus even for their proper phenomenological definition. I concluded that Mott insulators should simply be identified with the class of "ordinary" magnetic insulators, namely, the nominally periodic and stoichiometric compounds in which local moments and insulation both persist above the magnetic transition temperature T_N or T_C. (A small fraction of the known magnetic insulators are genuine ferromagnets; see I, p. 662.) Essentially this definition had long been adopted by most phenomenologically-oriented investigators, including Goodenough, Adler, and J. A. Wilson, but such a clear and simple identification was generally avoided by the more formal theoreticians. Anderson was one of the first to point out the connection between magnetic and Mott insulators, in his work on superexchange,[2] and other investigations pursuing this connection are discussed in I, §2. This was not the prevailing view, however.

One possible reason for this curious state of affairs was the historical accident that the antiferromagnetic nature of the favorite prototype material, NiO, was not firmly established until 1951, when the first detailed neutron scattering work was reported.[3] Another possible reason was that Slater's (1951) explanation[4] for the insulating gap, in terms of antiferromagnetic band theory, was seriously inadequate in several respects: (a) The finite-T prediction[5] was for a smooth decrease of gap and loss of sublattice magnetization, leading to an insulator → metal transition of mainly second-order character, rather than the observed moment-disordering (Néel or Curie) transition with essentially unchanged gap. (b) Subsequent calculations[6] for NiO produced an insulating (optical) gap far too small (~ 0.1 eV, rather than ~ 4 eV). (c) It offered no "natural" explanation for the Frenkel excitons (d^n optical absorption lines) seen in these materials, which are well described by crystal field theory. In any event, there was a conspicuous effort among theoreticians to try to explain the Mott insulating state without reference to magnetism, or at least to de-emphasize magnetic aspects as far as possible. The famous works of Hubbard[7] exemplify this trend, and helped to perpetuate it.

If one accepts the claim that NiO etc. have "localized" 3d electrons which form magnetic moments, and one desires to calculate features such as the Frenkel exciton spectrum, spin and orbital magnetic moment contributions, anisotropic g-factors, transferred hyperfine interactions, and superexchange couplings, it is well known that the ligand-field cluster model is appropriate. So, what is "the" Mott insulator problem? In general, this subject focuses on the interface between the "itinerant" (band-theoretic) and "localized" (cluster-model) descriptions, typical questions being (a) the origin of the insulating (optical) gap, (b) the detailed nature of the breakdown of conventional band theory, and the reasons for the success of the localized description, and (c) the nature of the general phase diagram, with respect to both experimental and formal model parameters. Accurate ab initio calculations are of course highly desirable, but a necessary prerequisite is a general conceptual framework. In Ref. 1 we have provided such a framework, which we shall now outline. Following this summary, we discuss some of the more recent works. Throughout this report we emphasize the mechanism by which the local-density approximation breaks down.

THEORY

The phenomenology of Mott insulators is quite rich, and no single formalism can encompass all of it. We shall concentrate on an unrestricted Hartree-Fock approach, which suffices to explain the insulating gap and the local-moment degrees of freedom. For pedagogical clarity we shall follow an "onion layer" approach, with increasing degrees of realism:

I.

The first stage considers the simple s-band Hubbard Hamiltonian for N sites and N electrons, and examines its eigenstates in Hartree-Fock (HF) approximation. (It is presumed that any "correlation corrections" to HF can be absorbed into effective values for the Hubbard parameters.) If U/W (W = bare bandwidth) is sufficiently large, there is a distinct self-consistent HF solution corresponding to each one of the 2^N possible spin configurations of the corresponding N-site Ising model. Such a solution is most easily achieved in Wannier representation (both conceptually and numerically; see I, Appendices A, B), by occupying an up-spin Wannier function on each up-spin site for the chosen Ising spin configuration, and similarly for the down-spins. Periodic boundary conditions are assumed so that Bloch periodicity is formally preserved, even though the "magnetic" unit cell is now the entire N-site model, for most of the possible Ising spin configurations.

The next step is transformation from Wannier to Bloch representation, as appropriate for the chosen magnetic unit cell. One need not know the detailed transformation matrix in order to see that all 2^N of these "Ising-Hartree-Fock" solutions must share some common features. For each spin configuration, the number of up-spin electrons is identical to the number of up-spin sites, thus the lower up-spin subband is exactly full. One can then show (using orthogonality and completeness, see Appendix B of I), that there is also an upper up-spin subband, empty and separated from the lower one by a gap of magnitude $\geq U - W$. Down-spins behave similarly, thus the familiar Bloch-Wilson explanation for insulation, in terms of filled and empty bands, is generalized from the simple antiferromagnetic configuration of Slater to all 2^N Ising spin configurations. At this most elementary level, the feature of insulation coexisting with local moments is now explained.

Each occupied Wannier orbital has "tails" extending onto neighboring sites, and its detailed form depends on the local spin environment. For parallel-spin neighbors, orthogonality requires that each of the Wannier functions must have a node between these sites, while for opposite-spin neighbors such nodes are missing. By considering the total HF energies for the various spin configurations, and examining their differences, one can thereby obtain an effective Heisenberg spin Hamiltonian. This procedure is fully consistent with the superexchange theory for a pair of magnetic ions in a nonmagnetic matrix,[2] and it extends that theory to the "concentrated" case of N magnetic ions. In general, however, the differences of the total HF energies are so small that direct evaluation is less sensible than a perturbation-theoretic approach. For d electrons, the total Heisenberg coupling also has some contributions not described by the present HF approach, but contained in a systematic perturbation theory. We return to this point below.

We emphasize that the ground state and its optical-gap excitations are described here in a conventional band-theoretic manner. That is, the one-electron excitations have not required a treatment of correlations different from that of the ground state. Most of the so-called "strong-correlation" problem is handled simply by considering the appropriate spin-polarized solution of the HF equations. A remaining (and energetically much weaker) part of the strong-correlation problem is manifested in the local-moment degrees of freedom and their Heisenberg spin coupling, which are treated here by means of <u>strong orbital</u> <u>rearrangement</u>. This is simply the fact that the self-consistent Ising-Hartree-Fock orbitals differ for different Ising spin configurations. This feature is quite simple and intuitive when viewed in terms of the appropriate Wannier representation. This approach also serves to explain the "localized electron" phenomenology.

The qualitative features of the general (U,T) phase diagram, as exemplified by V_2O_3 and $Ni(S,Se)_2$, can be understood already at this elementary level. This involves comparison of the single-spin-reversal energy ($\sim zJ \sim 2zt^2/U$, $z =$ nearest-neighbor coordination number) with the one-electron gap-crossing excitation energy ($\sim U - W$) for a fixed spin configuration, as well as comparing the resulting free energies with that of the ordinary metallic state. The role of dimensionality, crystal structure, and the resulting band structure is also clarified. See I, §4, for details.

II.

In the second stage of refinement, we replace the s orbitals by 3d orbitals exposed to the appropriate crystal field. The near-neighbor hopping parameters t are now replaced by a rather large set of parameters. Fortunately, most of these are very small,[8] and for present purposes only a few will need to be treated explicitly. The one-site Coulomb and exchange interactions among the 3d electrons, which arise from the tight-binding limit of HF theory, must be treated carefully. These interactions can be parametrized in several ways, for example by effective Slater integrals (F_0, F_2, F_4) or by the Racah parameters (A, B, C). However, the somewhat cruder parametrization of Kanamori[9] is convenient and will often suffice. Let index μ label the five spatial crystal-field orbital eigenstates for a single 3d electron. The Kanamori parameters are defined by

$$U = <\mu\mu|v|\mu\mu>, \qquad U' = <\mu\mu'|v|\mu\mu'> ,$$

$$J = <\mu\mu'|v|\mu'\mu>, \qquad \mu' \neq \mu ,$$

(1)

where v is the appropriate screened Coulomb interaction. The U' and J differ for different pairs $\mu\mu'$, thus some averaging is implied. Physically, this averaging destroys the second Hund rule (coupling to maximum allowed L), but this is often quite acceptable for the 3d Mott insulators because L is already strongly quenched by the crystal field. (This parametrization is, of course, not acceptable for the Frenkel exciton spectrum.) Simple averaging for U' and J leads to the useful relation $U - U' = 2J$.

The insulating gap is now explained as follows, using NiO as the example. Here the crystal-field orbital eigenstates consist of a lower triplet (xy, xz, yz) labeled t_{2g}, and an upper doublet (x^2-y^2, $z^2-1/3r^2$) labeled e_g. Ignoring the hopping parameters t for the moment, as well as spin-orbit coupling, we populate each site with eight 3d electrons in the strong-crystal-field configuration $e_g\uparrow^2$ $t_{2g}\uparrow^3 t_{2g}\downarrow^3$. (Here "up" and "down" now mean the majority and minority spins for each site, according to the assigned Ising configuration.) Using the Kanamori parameters (1) to determine the HF potential energies, it is easily seen that the assigned configuration is the self-consistent HF ground state. The empty $e_g\downarrow$ orbitals are raised above the highest occupied orbitals (the $t_{2g}\downarrow$'s) by $\Delta_{CF} + U'$ $- J \approx 4.9$ eV. (The new phenomenological values are $U' \approx 4.5$ eV, $J \approx 0.7$ eV as discussed below. The crystal-field splitting $\Delta_{CF} = 10$ Dq is 1.1 eV.) Among the many near-neighbor transfer integrals (hopping parameters), only three are obviously significant. These are $t_{ee} \equiv t(e_g, e_g, 100)$, $t_{tt} \equiv t(t_{2g}, t_{2g}, 1/2\ 1/2\ 0)$, and $t_{et} \equiv t(e_g, t_{2g}, 1/2\ 1/2\ 0)$. (See I, pp 719-720.) Ignoring the last of these, one finds effective e_g and t_{2g} bandwidths ($W_e \approx 2.0$ eV and $W_t \approx 1.6$ eV) which reduce the gap somewhat.

The effect of t_{et} is particularly interesting, because this tends to mix the orbital symmetries, i.e., to partially occupy $e_g\downarrow$ and partially empty $t_{2g}\downarrow$. Consider for a moment the limiting case of complete orbital-symmetry mixing, where the $\downarrow$ orbitals are all equally populated. In this case the $e_g\downarrow$-$t_{2g}\downarrow$ centroid splitting would be only $\Delta_{CF} = 1.1$ eV, and the subband widths W_e, W_t would then suffice to eliminate the gap. (This statement is appropriate for $T > T_N$. In the ground-state antiferromagnetic spin configuration W_e is greatly reduced, thus a small gap ~ 0.1 eV is found.[6,10-14]) This gap-closure does not happen in NiO because the symmetry-mixing transfer integral $t_{et} \approx 0.07$ eV is very much smaller than the "bare" gap of 4.9 eV, hence the occupation of $e_g\downarrow$ is only $\sim z(t_{et}/\text{bare gap})^2 \sim 10^{-3}$, and similarly for the emptiness of $t_{2g}\downarrow$. We conclude that strong orbital polarization is essential for the large gap.

Consider again the limit of complete orbital-symmetry mixing, with all $\downarrow$ orbitals equally populated, and likewise for the $\uparrow$ orbitals. The HF potential then produces only a simple spin splitting,

$$V_{HF}(\downarrow) - V_{HF}(\uparrow) = (n_\uparrow - n_\downarrow)(U + 4J)/5 , \qquad (2)$$

where n_σ is the number of σ-spin electrons per site. Therefore, as the degree of orbital polarization varies, the physical situation interpolates between this ordinary spin splitting and the Mott or "atomic HF" type of splitting between empty and filled orbitals. These two limits coincide only in cases like Mn^{2+} and Fe^{3+}, where $n_\uparrow = 5$ and $n_\downarrow = 0$. It is also important to note that the situation (2) is generated, in effect, whenever one spherically averages within the muffin-tin potential (I, Appendix D), and that this physical picture also coincides with that of the LSDA.

There are two important lessons to be learned here, especially for those band theorists who are devoted to the local-spin-density approximation (LSDA). First, the different-orbital ($\mu' \neq \mu$) exchange parameter $J \approx 0.7$ eV is very much smaller than the self-exchange ($\mu' = \mu$) parameter $U \approx 5.8$ eV. Use of the LSDA amounts, as we have just seen, to averaging over these enormously-different quantities. It thereby misses the main point of the Mott picture, namely, that the HF potential of an unoccupied orbital lies above that of an occupied orbital by an amount of order U. This point has been repeatedly confirmed by LSDA band calculations, which obtain little or no insulating gap for NiO.[6,10-14]

Having just attacked the LSDA, the second lesson is to understand how it (and the non-polarized LDA) can work so well for so many materials. A partial explanation is that in a metallic state the metallic screening (due mainly to RPA-type correlations) greatly reduces the effective U parameter, so the exchange-averaging just mentioned becomes a less serious approximation. There is, however, another consideration of major importance. The materials where LDA or LSDA work well typically have normal bandwidths, i.e., they are usually not "narrow-band" materials. One of the consequences is that the symmetry-mixing transfer matrix elements (analogs of our t_{et}) are generally much larger than for NiO. Thus, with larger "t_{et}" and smaller U, the occupied Bloch orbitals tend to nearly equally populate the various 3d orbitals, and likewise for the other $n\ell$

orbitals. In other words the _itinerancy_ (hopping between sites) tends to average the individual Wannier orbital occupations, and to thereby produce the same physical effect as the exchange-averaging which is implicit in the LDA or LSDA. We believe that this is the main reason why the LSDA often works well, in spite of its serious failure for NiO. We have also suggested (I, p. 749) that NiS is an intermediate case, with partial mixing of the minority-spin orbital symmetries. Obviously, a proper treatment of this orbital polarization issue requires use of the Kanamori parametrization, or something similar, as discussed below. We shall also comment further on the 3d transition metals below.

Another facet of this procedure is illustrated by CoO, where the zeroth-order 3d configuration is $e_g\uparrow^2 t_{2g}\uparrow^3 t_{2g}\downarrow^2$. Now the classification of filled and empty orbitals fails to coincide with _either_ (a) the $\uparrow$, $\downarrow$ spin classification, which makes the MnO (3d5, $e_g\uparrow^2 t_{2g}\uparrow^3$) case rather simple, _or_ (b) the crystal-field labeling, which allowed us to simply distinguish between $e_g\downarrow$ (empty) and $t_{2g}\downarrow$ (occupied) for NiO. CoO thus provides an "acid test" for any proposed treatment of Mott insulators. We have shown in I that consistent use of the Kanamori parameters in HF approximation is sufficient to explain why CoO, NiO, and MnO are all very similar physically. The key feature again is strong orbital polarization, this time among the $t_{2g}\downarrow$ orbitals. Wakoh[15] has presented a simplified model calculation for CoO which illustrates the role of the competition between itinerancy and U in producing this orbital polarization and the resulting insulating gap. He found a rapid crossover in behavior as a function of U/W, consistent with the discussion in I (pp. 749, 750) for NiS. Also, the cases of V_2O_3 ($t_{2g}\uparrow^2$) and $Ti_2O_3(t_{2g}\uparrow)$ have been studied within this framework, in much detail.[16] These cases are even more complex, with one electron per site participating in a nonmagnetic covalent bond.

Because there has been so much confusion about these issues, we shall now derive the Kanamori-parameter formula for the tight-binding limit of unrestricted HF theory. Let each (3d) Bloch function be expanded in a suitable Wannier basis,

$$\psi_{nk\sigma}(r) = N^{-1/2} \sum_{j\mu} C^{j\mu}_{nk\sigma} \phi_{\mu\sigma}(r-R_j) \ , \tag{3}$$

where n is a band index and μ can now refer to any representation of the 3d spatial functions. In a simple nonmagnetic state we would have

$$C^{j\mu}_{nk\sigma} = C^{\mu}_{nk\sigma} \ exp(ik\bullet R_j) \ ,$$

but we need the present form in order to describe general Ising-Hartree-Fock states. Consider the effect of HF exchange acting on one of these Bloch functions,

$$<r|V_x|\psi_{nk\sigma}> = {\sum_{n'k'}}' \int \psi^*_{n'k'\sigma}(r') \psi_{n'k'\sigma}(r) \psi_{nk\sigma}(r') v(r,r')d^3r' \ , \tag{4}$$

where Σ' means summation over occupied states, and $v(r,r')$ represents the screened Coulomb interaction. Expanding each of the Bloch functions in the Wannier basis and keeping only the one-site (Rj" = Rj' = Rj) terms, this becomes

$$<r|V_x|\psi_{nk\sigma}> = N^{-1/2} \sum_{j\mu} C^{j\mu}_{nk\sigma} \sum_{\mu'\mu''} P^{j\sigma}_{\mu'\mu''} <r\phi_{\mu'j\sigma}|v|\phi_{\mu''j\sigma}\phi_{\mu j\sigma}> \ , \tag{5}$$

where

$$P^{j\sigma}_{\mu'\mu''} = \frac{1}{N} {\sum_{n'k'}}' \left(C^{j\mu'}_{nk\sigma}\right)^* C^{j\mu''}_{nk\sigma} \tag{6}$$

is the one-body density matrix, and $\phi_{\mu j\sigma} = \phi_{\mu\sigma}(r-R_j)$. At this point we choose a new Wannier basis which diagonalizes the density matrix,

$$P^{j\sigma}_{\mu\mu'} \rightarrow P^{j\sigma}_{\mu} \delta_{\mu\mu'} \ ,$$

where we assume that the same spatial representation suffices for both σ's (and all j's). The matrix elements in (5) can now be replaced by their Kanamori-parameter equivalents, giving

$$V_X \Psi_{nk\sigma} = N^{-1/2} \sum_{j\mu} C^{j\mu}_{nk\sigma} \left\{ U P^{j\sigma}_{\mu} + J \sum_{\mu' \neq \mu} P^{j\sigma}_{\mu'} \right\} \Phi_{\mu j\sigma} \ . \tag{7}$$

Similar treatment for the direct potential term gives

$$V_D \Psi_{nk\sigma} = N^{-1/2} \sum_{j\mu} C^{j\mu}_{nk\sigma} \left\{ \sum_{\sigma'} \left[U P^{j\sigma'}_{\mu} + U' \sum_{\mu' \neq \mu} P^{j\sigma'}_{\mu'} \right] \right\} \Phi_{\mu j\sigma} \ , \tag{8}$$

therefore the total (intra-atomic) HF potential acting on each $\Phi_{\mu j\sigma}$ is

$$V_{HF}(\mu j\sigma) = U P^{j\bar{\sigma}}_{\mu} + U' \sum_{\mu' \neq \mu} (P^{j\sigma}_{\mu'} + P^{j\bar{\sigma}}_{\mu'}) - J \sum_{\mu' \neq \mu} P^{j\sigma}_{\mu'} \ , \tag{9}$$

where $\bar{\sigma} = -\sigma$.

For NiO the appropriate μ basis is obviously the one used above. For CoO, however, the optimum basis involves a non-trivial rotation within the t_{2g} subspace, for which the main "driving force" is spin-orbit coupling. Its physical consequence is an orbital contribution to the magnetic moment,[17] since the present degeneracy within the t_{2g} subspace permits a partial unquenching of the orbital angular momentum. (For an accurate treatment here, one should of course use Racah instead of Kanamori parameters.) In these examples the form of the orbital polarization is determined by one-electron effects (crystal-field splitting and spin-orbit coupling), while the intra-atomic interactions enhance this polarization and also add spin polarization.

III.

Atomic orbitals other than 3d must be considered. We start with the ligand p orbitals, the oxygen 2p's for NiO. Experiments[18] indicate that the centroid of the occupied 3d levels (or at least the $t_{2g} \downarrow$ level) lies above that of the 2p's. The "3d" functions of stage II are therefore really the antibonding 2p-3p hybrids of the ligand-field cluster model, or, rather, their band-theoretic analogs. It is well known that this covalency leads to some reduction of the effective intra-atomic interactions (the Slater, Racah, or Kanamori parameters, and the spin-orbit parameter λ_{so}), to the transferred hyperfine interaction, and also to the Goodenough-Kanamori systematics[2] for the various superexchange interactions.

What is not so well appreciated is that the degree of 2p-3d covalency is easily perturbed, and constitutes an electronic "soft mode" for the system. For example, much of the observed crystal-field splitting arises from covalency.[19] Removal of a "3d" electron, with its Coulomb repulsion, allows the remaining occupied 3d levels to fall closer to the 2p centroid, thereby increasing the covalency and approximately doubling the crystal-field splitting for the Ni^{3+} ion. (The latter consequence is well known to inorganic chemists.[2]) One may question whether part of this large increase is due to polaronic lattice distortion,[14] and may thus not be realized in "sudden" phenomena like photoemission. However, analysis of XPS data does indicate an increase of about this magnitude (I, pp. 731-3).

Actually, all of the ligand-anion hybridization channels (oxygen 2s, 2p coupling to Ni 3d, 4s, 4p) respond simultaneously, and the 2p-4s channel may also be important here. One of the resulting effects is to produce a backflow of charge, so that when a "3d" electron is removed, the net change in the total Ni-ion charge is considerably less than unity.[20] This backflow is a major contribution to the (solid state) screening of the effective U parameter. It also helps to explain why sulphur, which is more covalent than oxygen, leads to a smaller U. (This is an important ingredient of the chemists rule that sulphur is "more polarizable" than oxygen.) And similarly for the bulk dielectric polarizability. We suggest that the dominant effect here is not intratomic polarization, as usually assumed, but rather that this comes from changes in the covalency. There is some independent evidence for this.[21]

In I we examined a wide variety of NiO data to try to extract empirical values for the various t's (§5.5), as well as for the Kanamori and Racah parameters (§5.6.1, §5.7, Appendix C). Straightforward interpretations typically led to t values which were mutually inconsistent, and also in clear disagreement with the band-theoretic values of Mattheiss.[8] We found, however, that these disparate t values could be reconciled by considering correlation corrections arising from these covalency degrees of freedom (I, Appendix E).

IV.

We come at last to consideration of the valence (2p) and conduction (4s, 4p) bands of NiO. These bands are undoubtedly well described by conventional local-density band theory. A major problem, however, is the placement of the 3d level (say, for the removal energy of $t_{2g}\downarrow$ from $3d^8$) with respect to these bands. Our preference has been to treat this parameter empirically (as well as the Kanamori, Racah, and Δ_{CF} parameters), based on skepticism about the reliability of any of the existing ab initio methods. This situation is basically unchanged, except that there is now much more phenomenological information available as a result of fitting data with a detailed spectroscopy theory. We comment on this below.

There are two major aspects of Mott-insulator phenomenology that the present unrestricted HF approach is unable to handle. Within the standard crystal-field model, the Frenkel exciton states typically have multi-determinant wavefunctions, i.e. they involve strong configuration mixing. Inclusion within the present scheme would thus require generalization to the multi-determinant version of HF theory.[22] Although this would be computationally difficult (and not worth the effort), the extention is conceptually straightforward. In this sense the present scheme is consistent with the existence of these excitons, in contrast to LSDA models.

The other problem is that the present scheme really deals only with Ising degrees of freedom for the local moments. It does not account for the 2J + 1 states of actual local moments, nor for the more subtle aspects of their effective spin couplings (bi-quadratic exchange, many-site couplings, various spin-orbit coupling effects). In I (§6.5) we presented a form of many-body perturbation theory which can deal systematically with all of these complexities, as well as with all of the various known higher-order contributions to the conventional Heisenberg interaction. This is a linked-cluster formalism, hence it is not bothered at all by having a macroscopic number, N, of interacting moments. We claim that this formalism has provided the first clean and general resolution of the old "non-orthogonality catastrophe," which has plagued previous attempts to extend the usual two-magnetic-site superexchange theory to macroscopic N.

DISCUSSION

We now discuss some of the more recent developments, particularly for NiO. As implied above, we shall see that the most widespread type of confusion remaining today is that concerning Stage II above, namely, the issue of strong orbital polarization.

The most important new experimental information is undoubtedly the single-crystal BIS data[23] for NiO, combined with PES, which shows a very prominent BIS peak 5.6 eV above the sharp first peak in PES. The low energy, sharpness, and large magnitude of this BIS peak all combine to label this unambiguously as 3d addition, which must be the addition of $e_g\downarrow$ to the ground state $e_g\uparrow^2 t_{2g}\uparrow^3 t_{2g}\downarrow^3$. The threshold PES peak is likewise clearly due to removal of a 3d electron, as this peak accurately displays the multiplet structure predicted by crystal-field theory (I, pp. 731-3). This assignment is quite secure because the feature of multiplet structure agreement is equally true for CoO and MnO,[1] which have quite different multiplet signatures.[18] (Note that good-quality CF fits to this data require

$$\Delta_{CF}^{3+} \sim 2\,\Delta_{CF}^{2+},$$

consistent with general chemical experience, but overlooked in much of the Mott-insulator literature.) This 5.6 eV energy separation of the peaks is therefore a direct measurement of the "Mott" form of U, as defined by the d^n ground states; $U_M = E(d^9, {}^2E) + E(d^7, {}^4T_1) - 2E(d^8, {}^3A_2)$. Standard crystal-field theory[24] gives $U_M = A + 4B + \Delta_{CF}$. Using CF parameter values for the d^8 Frenkel excitons of NiO (B = 0.10 eV, C = 0.42 eV, Δ_{CF} = 10 Dq = 1.1 eV),[25] we thus find $U^0_M = U_M - \Delta_{CF} = 4.5$ eV, $A = F_0 - 7/5 C = 4.1$ eV, $F_0 = 4.7$ eV, and also the Kanamori parameter values,

$$U_K = A + 4B + 3C = 5.8 \; eV \; ,$$

$$J = 5/2\,B + C = 0.67 \; eV \; , \tag{10}$$

$$U'_K = U_K - 2J = 4.5 \; eV \; .$$

These "effective" values incorporate <u>all</u> screening/correlation effects phenomenlogically. Parametrization schemes which treat some of these effects explicitly should of course produce somewhat different values.

These values are far smaller than the rough concensus of a decade ago,[1] that $U^0_M \sim 8$ eV (and thus $U_K \sim 10$ eV). Nevertheless, they are still large compared to the subband widths W_e, $W_t \lesssim 2$ eV, so they still suffice to justify the preceeding theoretical picture. For comparison, we note that the spectroscopy of free Ni ions gives $U^{free}_M = E(d^9) + E(d^7) - 2E(d^8) = 18.0$ eV, and from the corresponding Racah expressions and d^8 parameters one finds $F_0 = 18.6$ eV. In going from free ions to NiO the effective F_0 therefore decreases by a factor of 4, in sharp contrast to the Racah B and C parameters (corresponding to Slater's F^2 and F^4) which decrease only $\sim 20\%$.

It was formerly thought that the onset of $2p \rightarrow 3d$ transitions would necessarily occur considerably above the optical gap edge of 3.8 eV, but with the present smaller U, together with the lowered position of the occupied 3d centroid (see below), this is no longer so. Both $2p \rightarrow 3d$ and intersite $3d(d^8 + d^8 \rightarrow d^7 + d^9$, especially first-neighbor) transitions are now reasonable candidates for the onset of the strong optical absorption. We know of no definitive evidence for either of these assignments, and it could be that both are contributing here. (This would correspond to the "intermediate case" in the classification scheme of Zaanen, Sawatzky, and Allen,[26] which is also their assignment for NiO.) However, a strong case has been made for the 2p-3d assignment, on the basis of the gap systematics for several series of compounds.[27] Optical data for MgO[28] and most band calculations[8,10,14,28-31] place the $2p \rightarrow 4s$ edge far above the gap, at around 8 eV. A strong-correlation band calculation has also placed the $3d \rightarrow 4s$ edge at around 8 eV.[31]

Photoemission spectra for NiO show a satellite 7 eV beyond the prominent first peak, and both features display strong Ni $3p \rightarrow 3d$ resonance behavior.[32] Sophisticated calculations using cluster[33] and single-impurity[34] models have fitted this data with reasonable parameters. These calculations consider only nickel 3d and oxygen 2p (and sometimes $2s$[33]) orbitals. Another process, 2p-4s shakeup accompanying 3d removal, could also be contributing satellite structure in this energy region, but other cluster calculations suggest that this intensity should be quite weak.[35]

The development of a unified spectroscopy theory,[33,34,36,37] with fitting parameters having nearly the same values for a number of different spectroscopies, is certainly a major achievement. As used to date, however, this theory has an unfortunate feature. It uses a free-ion basis for the d^n configurations, and treats all of the covalent hybridization explicitly. This contrasts with the Anderson approach[2] which employs "bonding" and "antibonding" basis functions, so that much of the ground state 2p-3d hybridization is incorporated from the outset. The Anderson description for the physical ground state of Ni^{2+} in NiO is simply d_{AB}^8 where d_{AB} signifies a 2p-3d antibonding orbital, while the new theory describes the same state as $\alpha|d^8\rangle +$

$\beta|d^9\underline{L}>$ + $\gamma|d^{10}\underline{L}^2>$ ($\underline{L}$ = ligand hole = $2p^{-1}$ or $2s^{-1}$). This free-ion basis is formally advantageous in several respects, but it is physically rather opaque. One knows, for example, that the Frenkel excitons are well described by the phenomenological crystal-field multiplet theory,[24] so there is evidently a high degree of coherence between the various free-ion components. In the recent work this coherence aspect has too often been ignored or underrated. Because the initial PES features of NiO, CoO, and MnO are all well described as CF multiplet structures arising from sudden removal of a d electron, it is quite clear that a $(d'_{AB})^7$ assignment is physically appropriate here for NiO (d'_{AB} meaning antibonding for the Ni^{3+} ion). Also, Larsson[35,38] has shown that the satellites in the core-state XPS of Ni compounds are well described as $(d'_{AB})^9\underline{L}$ or $(d'_{AB})^7\underline{L}$. By analogy, then, the valence-band satellite should be $(d'_{AB})^{8*}\underline{L}$ (* denoting possible multiplet excitation), where this state (or group of states) arises from 2p-3d shakeup following 3d emission. An effort to interpret the new results "coherently" á la Larsson[35,38] would thus be welcome.

We now consider the degree of progress towards ab initio calculations, first examining the band calculations. Self-consistent calculations[10,14,29-31,39] place the 3d centroid several eV lower than the non-self-consistent calculation of Mattheiss.[8] Spin-polarized calculations[10-14,31] have basically reconfirmed the old result of Wilson,[6] that the antiferromagnetic ground state configuration gives only a small energy gap for NiO. (The LSDA antiferromagnetic gap of Williams and co-workers has been reported as 0.3 eV[12] and 0.6 eV.[14]) Not only is this gap an order of magnitude too small at T = 0, it also vanishes for $T > T_N$,[11] and there is no gap whatsoever for CoO, all in striking disagreement with experiment. Williams and co-workers[13] have explained away the CoO problem by appealing to the orbital-polarization idea of Refs. 1 and 15, and more recently[14] they claim to have cured the T = 0 gap magnitude problem by including a Mott-Hubbard U. Their picture thus seems to be evolving towards that of I, although there still remains an inconsistency discussed below. Other calculations coming close to the picture of I are those of Refs. 15, 31, and especially 16.

Several band-theoretic efforts have managed to go beyond these local-density results. Kübler and Williams[14] and Norman and Freeman[40] and have each done supercell calculations, comparing energies of d^7, d^8, and d^9 site-occupations, to determine the Mott-Hubbard U. The results are 4 eV and 7.9 eV respectively. The proper quantity to compare the first of these values with is U^0_M, shown above to be 4.5 eV, thus the Kübler-Williams result is close. This U^0_M is the quantity appropriate for the Hubbard model, where all effects of the covalent hybridization are assumed to the absorbed into the parameters U and t. In contrast, the Norman-Freeman calculation did not allow the 3d orbitals to hybridize at all. Its output therefore corresponds to the U of the cluster and Anderson-impurity models employed in the spectroscopy theories,[33,34,36,37] where covalent hybridization is treated explicitly. We presume that spherical averaging was employed, whereby the calculated value actually represents F_0. The resulting Racah A is therefore 7.3 eV, to be compared with the spectroscopic value of about 6.5 eV.[34]

Clearly, both of these "ab initio" U calculations have done quite well. Further studies to test and perhaps improve the reliability of these supercell techniques would be welcome, and indeed there has been work on this for Ce ions.[41] It should also be interesting to carry out similar (d^7, d^8, d^9) calculations for a NiO_6 cluster, using the methods of theoretical chemistry[42] and a large basis set. Actually, bulk NiO has been treated in essentially this manner by Kunz.[31] He has first calculated the spin-polarized antiferromagnetic band structure in the full nonlocal Hartree-Fock approximation, using a local-orbital-basis method. He has then calculated correlation energy corrections for addition or removal of electrons in the various Bloch states, using a combination of methods. This gave a U_M of 11.6 eV, and thus a U^0_M of about 10 eV. This is a very ambitious and commendable program, but it evidently needs further refinement. There is another direct ab initio calculation[43] for U, giving 3.3 eV for nickel, but the screening mechanism was chosen to represent a metal rather than a Mott insulator.

It would obviously be nice to have a simple but reliable recipe for estimating U from atomic data, bulk dielectric polarizability, electronegativities, etc., but none of the available prescriptions are satisfactory.[1,37] These recipes typically give too little screening, and thus overestimate U. The foregoing evidence indicates that the dynamical responses of the covalent hybridization channels need to be treated explicitly here.

Perhaps the main quantitative challenge remaining is to obtain a reliable computation procedure for locating the position of the occupied d levels (the d-electron removal energies) with respect to the remaining bands. The main problem here is to avoid double-counting for the exchange-correlation energy of the d electrons, assuming that LDA or LSDA has been used for the overall band structure. The method of Kunz[31] avoids the local-density approximation, and might therefore provide a reliable computation method.

A basic difficulty remains for the "local-density plus U" approach of Ref. 14. If applied naively, this method would produce a Mott gap between the occupied and unoccupied Bloch states of any material, with the "more localized" orbitals (i.e., the more lumpy Bloch functions such as 3d's) experiencing the larger splitting. What is it that distinguishes more ordinary materials, including the pure Fe and Ni metals, which physically have no such splitting, from NiO and CoO, which have a large unoccupied-occupied 3d splitting? The method itself provides no criterion for its applicability. Proper resolution of this problem requires appeal to the unrestricted form of nonlocal Hartree-Fock theory, as discussed in Stage II above. We reiterate some key features: (a) Spatial and spin labels of the various Wannier orbitals must be treated on an equal footing, because orbital polarization is typically just as important as spin polarization. (b) The potential-energy shift of each Bloch orbital is determined by its relative weighting of each of its Wannier-orbital components, together with the self-consistent total occupation probabilities of all of these components, i.e., the overall spin-orbital polarization. The resulting energy shifts may or may not produce a splitting between the occupied and unoccupied Bloch states (I,§4.2.1). (c) The Mott-Hubbard ratio U/W must typically be rather large ($\gtrsim 1$) to produce strong orbital polarization, in view of the opposing effect of itinerancy, although the crystal-field splitting also contributes to the polarization. Thus for wide-band materials, and also for metallic materials where U is more strongly screened, the mechanism for Mott splitting is greatly weakened. Even then, however, effects of the state-dependent (Wannier-function dependent) potential (9) may remain observable. This has been demonstrated in studies of several 3d transition metals,[44] and also through the well known "gap problem" (underestimation of the insulating gap) for LDA band calculations of ionic insulators and noble-gas solids.[45]

We remark again, in this connection, that NiS is probably an intermediate case. With only moderate orbital polarization, the U which one would naively obtain from the very small gap (0.14 eV)[46] should be considerably smaller than the "true" U; see (2) above. This may be the origen of apparent inconsistencies in the phenomenology of this material.[47] (See I, p. 705 for a review of NiS.)

Because Mott insulators probably represent the most dramatic failure for the local-density band approach, we shall conclude by examining the present implications for the latter subject. Most of the older efforts to improve upon the local-density recipe (LDA or LSDA) have focused upon either (a) density-gradient corrections, (b) corrections with an explicit k-dependence, or (c) an nℓ (atomic-orbital)-dependent correction. Although some of these have significantly improved atomic calculations, their degree of success for the solid state is generally disappointing.[45] We have seen that what is most needed here is the nonlocal feature of the HF theory, so that the "magnetic" quantum numbers of the various atomic-orbital components, or rather their point-symmetry Wannier equivalents "μ" [as in (3)-(9) above] can be properly identified and utilized. The "self-energy correction" idea[45,48] comes closest to reproducing the present physics, but this scheme has encountered problems in transferring atomic experience to the solid state.[45] We can see two sources for these problems: (a) the issue of the degree of orbital polarization, and (b) solid-state effects which cause additional screening of the intra-atomic parameters.

A notable feature of essentially all attempts to improve upon local-density approximations is that they rely on numerical experience and theoretical methods (such as HF and many-body theory) which lie outside of the formal structure of the density-functional methodology. The features we have been focusing upon are further examples of this. Most of the available evidence for deficiencies of the LDA-LSDA (Mott insulators, 3d transition metals,[44] "localized" examples of the gap problem,[45] as well as the systematic errors in atomic and free-ion calculations[48]) confirms the physical reality of the present type of considerations, which are clearly related to the off-diagonal feature of the one-body density matrix $n(r,r')$. The Hohenberg-Kohn theorem[49] implies, however, that only its diagonal part, $n(r,r) \equiv n(r)$ is needed. How can this paradox be resolved? Beyond the fact of its existence, almost nothing is known about the form of the Hohenberg-Kohn energy functional $F[n(r)]$. Its dependence on $n(r)$ may be highly nonlocal, and the prospects for correctly determining its form seem very dim.[50] We therefore conclude that $F[n(r)]$ must somehow contain the present physics, presumably as the result of its nonlocality together with some global or integral identity for solutions of the many-electron Schrödinger equation. (It must be noted, however, that the exact density functional would not necessarily reproduce the one-electron excitation spectrum, which we are attempting to do.) Assuming this is so, there is no fundamental conflict between these approaches, and one should therefore be guided by practical considerations. The present approach is calculationally incomplete, but it does offer a general physical picture with prospects for continued quantitative refinement. It is also significant that the resolution of the gap problem for ordinary semiconductors has required explicit use of the nonlocality of a screened HF exchange potential,[51] consistent with the present picture.

I thank John Perdew for helpful discussions about density-functional theory. This work was supported by the U.S. Department of Energy.

REFERENCES

1. B. H. Brandow, Adv. Phys. 26, 651 (1977), referred to as I. A short account of this work is B. H. Brandow, Int. J. Quantum Chem. Symp. 10, 417 (1976).

2. P. W. Anderson, Solid State Phys. 14, 99 (1963).

3. C. G. Shull, W. A. Strauser, and E. O. Wollan, Phys. Rev. 83, 333 (1951).

4. J. C. Slater, Phys. Rev. 82, 538 (1951).

5. T. Matsubara and T. Yokota, Proc. Int. Conf. Theor. Phys., Kyoto, 1953 (Tokyo, Science Council Japan, 1954), p. 693.

6. T. M. Wilson, Int. J. Quantum Chem. Symp. 2, 269 (1968) and 3, 757 (1970).

7. J. Hubbard, Proc. Roy. Soc. A 276, 238 (1963) and 281, 401 (1964).

8. L. F. Mattheiss, Phys. Rev. B 5, 290 and 306 (1972).

9. J. Kanamori, Prog. Theoret. Phys. 30, 275 (1963).

10. O. K. Andersen, H. L. Skriver, H. Nohl, and B. Johannsson, Pure Appl. Chem. $\underline{52}$, 93 (1979).

11. T. Oguchi, K. Terakura, and A. R. Williams, Phys. Rev. B $\underline{28}$, 6443 (1983).

12. K. Terakura, A. R. Williams, T. Oguchi, and J. Kübler, Phys. Rev. Lett. $\underline{52}$, 1830 (1984).

13. K. Terakura, T. Oguchi, A. R. Williams, and J. Kübler, Phys. Rev. B $\underline{30}$, 4734 (1984).

14. J. Kübler and A. R. Williams, J. Mag. Mag. Mat. $\underline{54\text{-}57}$, 603 (1986).

15. S. Wakoh, J. Phys. F$\underline{7}$, L15 (1977).

16. J. Ashkenazi and M. Weger, J. de Physique $\underline{37}$, C4-189 (1976); C. Castellani, C. R. Natoli, and J. Ranninger, Phys. Rev. B $\underline{18}$, 4945, 4967, and 5001 (1978).

17. J. Kanamori, Prog. Theoret. Phys. $\underline{17}$, 177 (1957).

18. D. E. Eastman and J. L. Freeouf, Phys. Rev. Lett. $\underline{34}$, 395 (1975).

19. T. F. Soules, J. W. Richardson, and D. M. Vaught, Phys. Rev. B $\underline{3}$, 2186 (1971); A. J. H. Wachters and W. C. Nieupoort, Phys. Rev. B $\underline{5}$, 4291 (1972).

20. P. S. Bagus, U. I. Walgren, and J. Almlof, J. Chem. Phys. $\underline{64}$, 2324 (1976).

21. S. T. Pantelides, Phys. Rev. Lett. $\underline{35}$, 250 (1975).

22. J. Hinze, J. Chem. Phys. $\underline{59}$, 6424 (1974).

23. G. A. Sawatzky and J. W. Allen, Phys. Rev. Lett. $\underline{53}$, 2339 (1984).

24. J. S. Griffith, The Theory of Transition-Metal Ions (Cambridge, 1961).

25. D. Reinen, Bunsenges. Phys. Chem. Ber. $\underline{69}$, 82 (1965).

26. J. Zaanen, G. A. Sawatzky, and J. W. Allen, Phys. Rev. Lett. $\underline{55}$, 418 (1985).

27. S. Hüfner, Z. Phys. B $\underline{61}$, 135 (1985).

28. M. L. Cohen, P. J. Lin, D. M. Roessler, and W. C. Walker, Phys. Rev. $\underline{155}$, 992 (1967).

29. T. C. Collins, A. B. Kunz, and J. L. Ivey, Int. J. Quantum Chem. Symp. $\underline{9}$, 519 (1975).

30. J. Hugel, C. Carabatos, F. Bassani, and F. Casula, Phys. Rev. B $\underline{24}$, 5949 (1981); J. Hugel and C. Carabatos, J. Phys. C. $\underline{16}$, 6723 (1983).

31. A. B. Kunz, J. Phys. C. $\underline{14}$, L455 (1981), and Int. J. Quantum Chem. Symp. $\underline{15}$, 487 (1981).

32. S. J. Oh, J. W. Allen, I. Lindau, and C. J. Mikkelsen, Jr., Phys. Rev., B $\underline{26}$, 4845 (1982); M. R. Thuler, R. L. Benbow, and Z. Hurych, Phys. Rev. B $\underline{27}$, 2082 (1983).

33. A. Fujimori and F. Minami, Phys. Rev. B $\underline{30}$, 957 (1984).

34. J. Zaanen, Thesis (Groningen, 1986).

35. S. Larsson, Chem. Phys. Lett. 40, 362 (1976).

36. G. van der Laan, C. Westra, C. Haas, and G. A. Sawatzky, Phys. Rev. B 23, 4369 (1981).

37. J. Zaanen, C. Westra, and G. A. Sawatzky, Phys. Rev. B 33, 8060 (1986).

38. S. Larsson, Chem. Phys. Lett. 32, 401 (1975), Physica Scripta 16, 378 and 381 (1977), and 21, 558 (1980); S. Larsson and M. Braga, Chem. Phys. Lett. 48, 596 (1977).

39. A. B. Kunz and G. T. Surratt, Solid State Commun., 25, 9 (1978).

40. M. R. Norman and A. J. Freeman, Phys. Rev. B 33, 8896 (1986).

41. M. R. Norman, D. D. Koelling, A. J. Freeman, H. J. F. Jansen, B. I. Min, T. Oguchi, and L. Ye, Phys. Rev. Lett. 53, 1673 (1984); B. I. Min, H. J. F. Jansen, T. Oguchi, and A. J. Freeman, Phys. Rev. B 33, 8005 (1986); A. K. McMahan and R. M. Martin, this proceedings.

42. See for example E. Miyoshi, T. Takada, S. Obara, H. Kashiwagi, and K. Ohno, Int. J. Quantum Chem. 19, 451 (1981).

43. B. N. Cox, M. A. Coulthard, and P. Lloyd, J. Phys. F 4, 807 (1974).

44. L. Hodges, H. Ehrenreich, and N. D. Lang, Phys. Rev. 152, 505 (1966); S. Wakoh and J. Yamashita, J. Phys. Soc. Japan 35, 1394 (1973); J. F. Cooke, J. W. Lynn, and H. L. Davis, Phys. Rev. B 21, 4118 (1980); A. Liebsch, Phys. Rev. Lett. 43, 1431 (1979) and Phys. Rev. B 23, 5203 (1981); G. Treglia, F. Ducastelle, and D. Spanjaard, J. Physique 43, 341 (1982); O. Bisi, C. Calandra, U. del Pennino, P. Sassaroli, and S. Valeri, Phys. Rev. B 30, 5696 (1984); L. C. Davis, J. Appl. Phys. 59 (6), R25 (1986).

45. J. P. Perdew, Int. J. Quantum Chem. Symp. 19, 497 (1986), and references therein.

46. A. S. Barker and J. P. Remeika, Phys. Rev. B 10, 987 (1974).

47. J. W. Allen, private communication.

48. J. P. Perdew and A. Zunger, Phys. Rev. B 23, 5048 (1981).

49. P. Hohenberg and W. Kohn, Phys. Rev. 136, B864 (1964).

50. E. H. Lieb, Int. J. Quantum Chem. 14, 243 (1983).

51. R. W. Godby, M. Schluter, and L. J. Sham, Phys. Rev. Lett. 56, 2415 (1986) and Phys. Rev. B 35, 4170 (1987). See also M. S. Hybertson and S. G. Louie, Phys. Rev. Lett. 55, 1418 (1985) and Phys. Rev. B 34, 5390 (1986).

QUALITATIVE CONSIDERATIONS ON TRANSITION METAL COMPOUNDS

C. Haas

Laboratory of Inorganic Chemistry, Materials Science Centre
Nijenborgh 16, 9747 AG Groningen
The Netherlands

1. NEGATIVE HUBBARD U

In many semiquantitative considerations on transition metal compounds, the Coulomb interaction between two electrons at one atom is represented by the so-called "Hubbard U". U is also the energy necessary for charge transfer, i.e. the energy of the chemical reaction $M(d^n)$ + $M(d^n) \rightarrow M(d^{n+1}) + M(d^{n-1})$. For a reaction between free ions $U_{at} = I_{n+1} - I_n$, where I_n is the ionization energy of an ion $M(d^n)$. For free ions $I_{n+1} > I_n$, so that $U > 0$. In a solid (or liquid) the interaction between two electrons on the same atom is screened: $U = U_{at} + \Delta U_{scr.}$, due to interaction with the surroundings. This screening is a result of the polarization of the neighbouring atoms, and has contributions from vibronic and electronic polarization. The screening leads to a lowering of U, so that $U < U_{at}$. This screening effect is generally quite large, and $\Delta U_{scr.}$ is usually of the same order as U_{at}. It is even possible that the net effect of screening is so large that $U < 0$ (overscreening). If this is the case, charge transfer $d^n + d^n \rightarrow d^{n+1} + d^{n-1}$ will occur spontaneously, and the state $d^{n+1} + d^{n-1}$ is the more stable one.

We briefly mention a number of cases with $U < 0$.
1) From Auger spectra it is possible to determine directly the value of U. Experimental data on Auger spectra of early transition metals Sc and Ti indicate a negative U for the interaction between electrons in the 3d bands.[1] Also from Auger spectra we find $U < 0$ for the interaction between holes in the valence band of $TiSe_2$ and $CrSe_2$[1].

2) The solid compound AuSe contains two types of gold ions, with different valencies: $Au^+(5d^{10})$ (with square planar coordination of Se ions) and $Au^{3+}(5d^8)$ (with octahedral coordination of Se ions).[2] This means that $Au^+(5d^{10}) + Au^{3+}(5d^8)$ has a lower energy than $2Au^{2+}(5d^9)$. The stabilization of the charge transfer state is presumably due to the polarization of the highly polarizable Se^{2-} ions.

3) The solid Cs_2SbCl_6 does not contain $Sb^{4+}(5s^1)$, but rather $Sb^{3+}(5s^2)+Sb^{5+}(5s^0)$ ions with different coordinations.[3] This structure is caused by the relative stability of the closed $5s^2$ shell (i.e. small value of U_{at}), combined with strong screening due to polarization of Cl^- ions.

4) Photoelectron spectra of the superconducting intercalation compound $SnTaS_2$ indicate that the valency of the Sn atoms fluctuates between the states $Sn^0(5s^25p^2)$ and $Sn^{2+}(5s^2)$.[4,5]

5) The layer compound TaS_2 has at low temperature a charge density wave state (CDW), in which there are different valencies on the Ta atoms: $Ta^{4+\delta}(5d^{1-\delta}) + Ta^{4-\delta}(5d^{1+\delta})$. Generally a CDW state is a partial charge transfer, stabilized by vibronic polarization which leads to clustering of metal atoms.[6]

6) The crystal structure of $BaBiO_3$ shows two types of Bi ions, which has been interpreted as evidence for a charge disproportionation into Bi^{3+} and Bi^{5+} ionic states.[7] Partial substitution of Bi by Pb leads to a metallic state with a high superconducting transition temperature T_c = 13 K. It has been suggested that the high T_c values (up to T_c = 93 K) reported for superconducting oxides like $Y_{1.2}Ba_{0.8}CuO_{4-\delta}$ are related to $Cu^+ - Cu^{3+}$ charge disproportionation.[8]

These examples show that small or negative values of U are quite common, and lead to interesting and anomalous physical properties.

2. MODEL HAMILTONIANS

For a detailed analysis of spectroscopic data one frequently uses model Hamiltonians. Well-known examples are the spin-Hamiltonians used in the interpretation of electron spin resonance data of transition metal or rare earth ions. Another example is the model Hamiltonian (Hubbard, Anderson Hamiltonian) used to analyse complicated photoelectron spectra, with satellites and the like due to electron correlation effects. These model Hamiltonians are very useful, but it is necessary to realize that they represent only a description, and not the whole truth. We make some remarks about the model. Hamiltonians used in photoelectron spectroscopy.

1) The model Hamiltonian is an approximation which describes interactions in a convenient and simple, but crude manner. Examples are the Hamiltonians with terms for hybridization $V_{mk}c_{m\sigma}^{+}c_{k\sigma}$, for electron-electron interaction $Un_{i\sigma}n_{-i\sigma}$, etc. It is well-known that Coulomb and exchange interactions cannot be represented exactly by U, and transfer not by a constant matrix element V_{mk} which is independent of the electronic state of all other electrons of the system. In many papers there is emphasis on the exact solution of the (model) Hamiltonian. However, this is equivalent to an approximate solution of an exact Hamiltonian. In the model Hamiltonian the severe approximations are hidden in the Hamiltonian, in many cases in an obscure way.

3) The parameters in the model Hamiltonian, if fitted to experimental data, are effective parameters which do not have the simple meaning which they have in the simple model system for which the Hamiltonian was derived.[9]

Consider as an example the so-called on-site Coulomb interaction between two electrons $Un_{i\sigma}n_{i-\sigma}$. For a single atom with 0, 1 or 2 electrons in a single orbital, U is the Coulomb repulsion $\langle e^{2}/r_{12}\rangle$ indeed. However, the value obtained from an analysis of photoelectron spectra in a solid is an effective U which can differ strongly from the atomic value by screening contributions. These screening effects are not on-site, and the effective U is not an on-site interaction.

2) In model Hamiltonians one usually assumes an orthogonal basis set of orbitals. This also implies that the orbitals are no longer localized on a single site, and if U represents the interaction between two electrons in orbitals of this orthogonal basis set, it is again not an on-site interaction. A consequence is that Coulomb interactions between different sites cannot be neglected; they are a necessary consequence of the orthogonalization of the basis set.

4) The use of effective parameters implies that the effect of other degrees of freedom is not treated explicitly, but rather in an effective manner. An example is that U represents the interaction reduced by screening. Therefore U can be considered as the interaction between two quasiparticles, each consisting of the electron and a polarization (screening) cloud. One should realize

that the creation operators C_n^+ create quasi particles, including the polarization cloud. Non-diagonal matrix elements between such quasi-particle states are now reduced by an overlap between the screening (polarization) clouds. These effects are well-known for vibronic coupling (Jahn-Teller, Franck Condon, Ham effect), but are usually not considered in theories on photoelectron spectra.[10,11)]

3. SPIN POLARONS IN HALF-METALLIC FERROMAGNETS

In many transition metal compounds there are fairly localized d-electrons and s-p electrons in broad energy bands. An example are the Ni 3d electrons in NiO which form localized magnetic moments and the O 2p electrons in broad energy bands. The interaction between these two types of electrons is a matter of great importance for the magnetic, electrical and optical properties of transition metal compounds.

An interesting class of materials are the halfmetallic ferromagnets, such as the Heusler alloys NiMnSb and PtMnSb.[12)] These compounds are strongly ferromagnetic. The exchange splitting between Mn 3d↑ and 3d↓ is quite large (~ 4 eV), and only the Mn 3d↑ states are occupied. There is an appreciable coupling between Mn 3d states and the Sb 5p states which form the broad valence band (width ~ 8 eV). A result is that the 5p↓-Sb band is completely occupied ($5p^3$↓ for every Sb), the 5p↑-Sb band is occupied by two electrons for each Sb($5p^2$↑). Therefore the Fermi level intersects the Sb 5p↑ band, and the Fermi level lies in a gap for electrons with spin ↓. Thus NiMnSb is metallic for spin ↑ electrons, it is semiconducting for spin ↓ electrons (half metallic ferromagnet). There is complete spin polarization of charge carriers (holes in the Sb-5p valence).

These band structure considerations are valid at T = 0 K. Of interest is what happens at higher temperature, in particular above the Curie temperature T_c, and in which way the spin polarization decreases with increasing temperature.

For the discussion of the electronic structure we consider a simple model Hamiltonian for a crystal with lattice sites $\vec{n}$, and a conduction electron orbital $\phi(\vec{r}-\vec{n})$ and a localized spin $\vec{S}_n$.

$$H = \sum_n \sum_\sigma \epsilon_o c_{n\sigma}^+ c_{n\sigma} + T_o \sum_n \sum_\sigma \sum_b c_{n\sigma}^+ c_{n+b,\sigma} + 2J \sum_n \vec{s}_n \vec{S}_n$$

ε_0 is the orbital energy, T_0 the transfer integral from site $\vec{n}$ to a neighbouring site $\vec{n}+\vec{b}$, and J is the exchange interaction between the conduction electron spin $\vec{s}_n$ with the local spin $\vec{S}_n$. There are two extreme cases:

a) weak exchange $JS << T_0$. There is an exchange splitting of the conduction band into two bands for electrons with spin parallel or antiparallel to the average magnetization M. This exchange splitting is proportional to M, and vanishes above T_c. This weak exchange model is usually employed to discuss properties of magnetic semiconductors[13], magnetic metals[14] and spin disorder scattering.[15]

b) strong exchange $JS >> T_0$. In this case the spin of an electron in an local orbital $\phi(\vec{r}-\vec{n})$ is strongly coupled to the local spin $\vec{S}_n$ at the same site.[16,17] There are at each site two local states $\varepsilon_n^+ = -J(S+1)$ and $\varepsilon_n^- = +JS$, with conduction electron spin parallel and antiparallel to the local spin $\vec{S}_n$. Thus there is a complete local exchange splitting which is independent of the magnetic ordering, and which persists also above T_c.

Two conduction bands ε_k^+ and ε_k^- are formed by transfer between the local states $\varepsilon_n^\pm$. However, the transfer integral now depends on the relative orientation of the spin on neighbouring sites $T(n, n+b) = T_0 \cos(\theta/2)$ (θ is the angle between $\vec{S}_n$ and $\vec{S}_{n+b}$).[18,19] This strong exchange model has been discussed by several authors, and has names as spin polaron, or local band model.

In the strong coupling model the average transfer $<T> = T_0 <\cos \theta/2>$ is responsible for the formation of the energy band, the fluctuations of the transfer are responsible for the spin disorder resistivity.

The case of the Heusler alloys NiMnSb is slightly more complicated, because the conduction electrons (or rather holes) are at Sb sites, and the localized spins at Mn sites. Therefore the exchange coupling is between the electron spin in an Sb 5p orbital and the spins of the six Mn atoms around this Sb atom.

Experimental data on the electrical resistivity and the spontaneous Hall effect (skew scattering and side jump) indicate that the strong coupling model is valid in NiMnSb.[20] A quantitative calculation of the spin disorder resistivity of NiMnSb, using values for J, T_0, m^* deduced from the band structure calculations, is in quantitative agreement with experiment.

The conclusion is that in NiMnSb the exchange coupling between the band electron and the local spin is strong (spin polaron). This might well be the case in many transition metal compounds (NiO).

REFERENCES

1) D.K.G. de Boer, C. Haas and G.A. Sawatzky, J. Phys. F.: Met. Phys. 14, 2769 (1984).

2) J.E. Cretier and G.A. Wiegers, Mat. Res. Bull., 8, 1427 (1973).

3) K. Prassides and P. Day, Inorg. Chem. 24, 1109 (1985).

4) R. Eppinga, G.A. Wiegers and C. Haas, Physica 105B, 174 (1981).

5) J. Dijkstra, C.F. van Bruggen, C. Haas and R.A. de Groot, to be published.

6) C. Haas, Current Topics in Materials Science, Vol. 3, 1 (1979).

7) T.M. Rice and L. Sneddor, Phys. Rev. Letters 47, 689 (1981).

8) M.K. Wu, J.R. Ashburn, C.J. Torng, P.H. Hor, R.L. Meng, L. Gao, Z.J. Huang, Y.Q. Wang and C.W. Chu, Phys. Rev. Letters 58, 908 (1987).

9) W. Beall Fowler and R.J. Elliott, Phys. Rev. B34, 5525 (1986).

10) J. Zaanen, C. Westra and G.A. Sawatzky, Phys. Rev. B33, 8060 (1986).

11) O. Gunnarsson and K. Schönhammer, Phys. Rev. B28, 4315 (1983).

12) R.A. de Groot, F.M. Mueller, P.G. van Engen and K.H.J. Buschow, Phys. Rev. Letters 50, 2024 (1983).

13) C. Haas, Crit. Rev. Solid State Sciences 1, 47 (1970).

14) T. Kasuya, Progress Theor. Phys. (Kyoto) 6, 45 (1956).

15) P.G. de Gennes and J. Friedel, J. Phys. Chem. Solids 4, 71 (1958).

16) E.L. Nagaev, Phys. Status Sol. b65, 11 (1974).

17) V. Korenman, J.L. Murray and R.E. Prange, Phys. Rev. B16, 4032 (1977).

18) P.W. Anderson and H. Hasegawa, Phys. Rev. 100, 675 (1955).

19) P.G. de Gennes, Phys. Rev. 118, 141 (1960).

20) M. Otto, H. Feil, R.A.M. van Woerden, J. Wijngaard, P.J. van der Valk, C.F. van Bruggen and C. Haas, to be published.

THE CHARACTER OF BAND GAPS IN TRANSITION METAL COMPOUNDS

G.A. Sawatzky

Laboratory of Solid State Physics, Materials Science Center

University of Groningen, Nijenborgh 16, 9747 AG Groningen

Perhaps the longest standing controversy in narrow band materials relates to the size and nature of the band gap in especially the late 3d transition metal compounds[1] which is of importance in understanding the systematics of band gaps[2], the closing of the gap in metallic systems3, the optical properties[4] and superexchange interactions[5] which involve virtual charge and spin excitations. Formally the band gap is determined by the minimum energy required to remove an electron from a system ($E^{N-1}-E^N$) plus the energy required to add one ($E^{N+1}-E^N$) where E^N is the ground state energy of the N electron system. This energy is equivalent to an electron-hole excitation in the N particle system to the lowest energy <u>dissociative</u> state i.e. the lowest energy of those states in which the electron and hole are uncorrelated. An important question then is what is the nature of the first ionized state and first electron affinity state in narrow band compounds? Obviously this is also an extremely relevant question in the new high temperature superconductors. For example the substitution of Ba for La in La_2CuO_4 or the addition of oxygen in $YBa_2Cu_3O_{6.5}$ must be charge compensated by holes in the valence band which then are the charge carriers. What is the nature of these holes - are they Cu 3d like or O2p like? In the dilute Ba/Sr substitution limit these holes are identical to the highest occupied states seen in photoemission of the pure unsubstituted compound. We have recently[6] tried to attain a qualitative physical answer to the above questions using an Anderson impurity Hamiltonian description for the late 3d TM compounds. In this way the various types and sizes of band gaps can be put into a framework described by a few parameters which at least have some physical meaning for most of us. We realize that the low energy scale properties cannot be described in this way since the translational symmetry is then certainly of importance, however perhaps we can say something about the nature of the quasi particles involved.

In this model we start with an ionic ansatz shown in figure 1 and ask for the possible states involving charge fluctuations. Basically there are two types which are important.

1. $d_i^n d_j^n \rightarrow d_i^{n-1} d_j^{n+1}$,   2. $d_i^n \rightarrow d_i^{n+1}\underline{L}$

where $\underline{L}$ denotes a hole in the anion p band and we have neglected the TM 4s band. The energies of these states are at

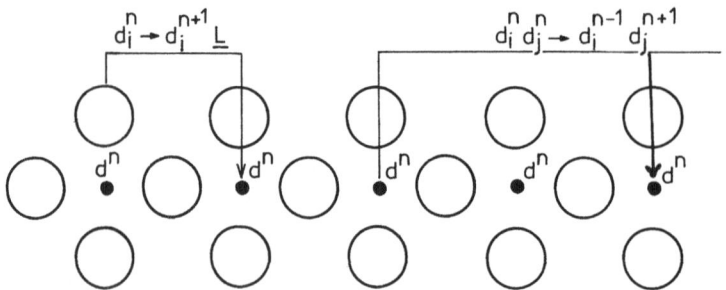

Figure 1. Representation of an ionic lattice consisting of TW ions (d^n)
and closed shell anions. The most important cahrge fluctuation
excitations are indicated.

$$E(1) = E(d^{n-1}) - E(d^n) + E(d^{n+1}) - E(d^n) \tag{1}$$

We define U as the mimimum possible such energy; in other words we take
the lowest energy (Hund's Rule) state in each case.

$$E(2) = E(d^{n+1}) - E(d^n) + \varepsilon_L = \Delta \tag{2}$$

where Δ is commonly called the charge transfer energy, which we can also
define as $\Delta = \varepsilon_d + \varepsilon_L$, in which case $E(d^{n-1}) - E(d^n) = U - \varepsilon_d$. Even in
the ionic picture there are two other quantities, the d band dispersional
width (w) and the anion p band dispersional width (W), which are of importance for the excited states. The states
$d_i^{n-1}d_j^{n+1}$ will in fact have a dispersional width of 2w because of the
translational symmetry of the TM ions. Also the excited states $d^{n+1}L_k$ will
have a dispersional width of W + w because of translational symmetry.

We are now in a position to draw a total energy diagram based on the
ionic ansatz as shown in Fig. 2 for $U \gg w$, $U > \Delta$, $\Delta > W$. In Fig. 2 we can
see the various types of band gaps which might occur. For $U > \Delta$ the gap is
of a charge transfer type and its magnitude is $\Delta - W/2$. So even for $U \to \infty$
we can get a metallic ground state if $\Delta < W/2$. Since generally $w \ll W$ these
materials will be p-type metals as for example CuS. For $\Delta > W/2$ the gap

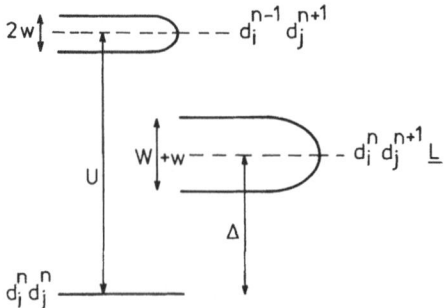

Figure 2. Total-energy-level diagram corresponding to an ionic ground state
and excitations as indicated in figure 1.

will scale as the anion electronegativity for a given cation and crystal structure. This is the case for the series $NiCl_2$, $NiBr_2$, NiI_2 with gaps of 4.7 eV, 3.5 eV and 1.7 eV respectively and a gap of zero for NiS.

For U < Δ we are in the Mott-Hubbard regime with a d-d gap for U > w, and a d band metal for U < w. It is generally accepted that the early 3d transition metal oxides belong to this regime.

We can put all of this information into a simple phase diagram shown in Fig. 3 which is a simplified version of the diagram including hybridization we have recently presented[6].

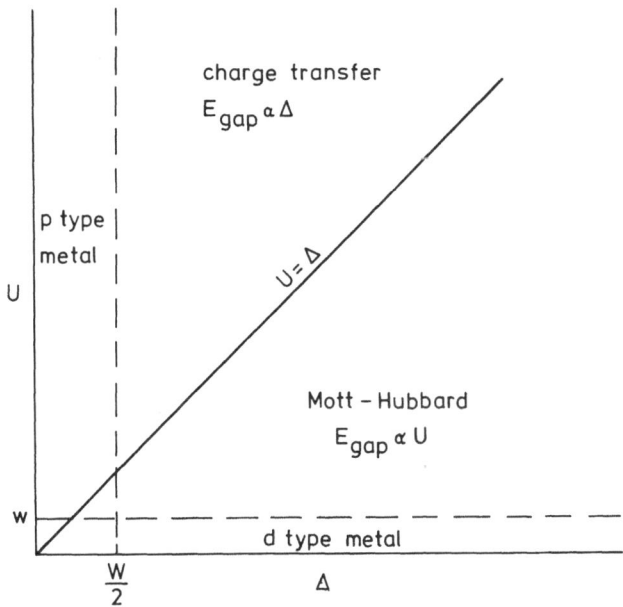

Figure 3. Simple phase diagram showing the various types of insulating and metallic states in transition metal compounds.

The nature of the first ionized and affinity states can be found by treating the TM ion as impurity in the lattice and determining the lowest energy N, N-1 and N+1 electron states. We can solve this problem with techniques developed by Gunnarsson and Schönhammer[7] for metallic systems. The important thing to note here is that if U is large (U > Δ) the energy level scheme for the N-1, N and N+1 electron system are all different resulting in a strong dependence of hybridization on the orbital occupation. This situation is shown in figure 4.

I want to especially draw attention to the N-1 particle states for U > Δ. For this situation the basis state energy of the d^{n-1} state lies considerably higher (farther from the fermi level) than the $d^n\underline{L}$ which is a ligand hole state. Switching on a sufficiently large hybridization now could cause a bound state to be pushed out of the bottom of the $d^n\underline{L}$ band in the impurity limit which would then be the lowest ionization state. This state would be a strongly mixed state of d^{n-1} and $d^n\underline{L}$ character which would form a narrow band in the translational symmetric case. An interpretation of photoemission, inverse photoemission, and optical data of NiO lead us to the conclusion that it is in this regime. We note that for W >> W the ground state, in contrast to this, is highly ionic. This is a nice example of how one must be careful in drawing conclusions concerning the degree of covalency or hybridization in the ground state from high energy spectroscopy. It is also interesting to note that Anderson in his theory of superexchange implicitly assumed that Δ >> U which according to our knowledge now of the late TM compounds is probably not correct. This does not change the basic physics of superexchange but does solve a problem related to the trend in Néel temperatures for the monoxides.

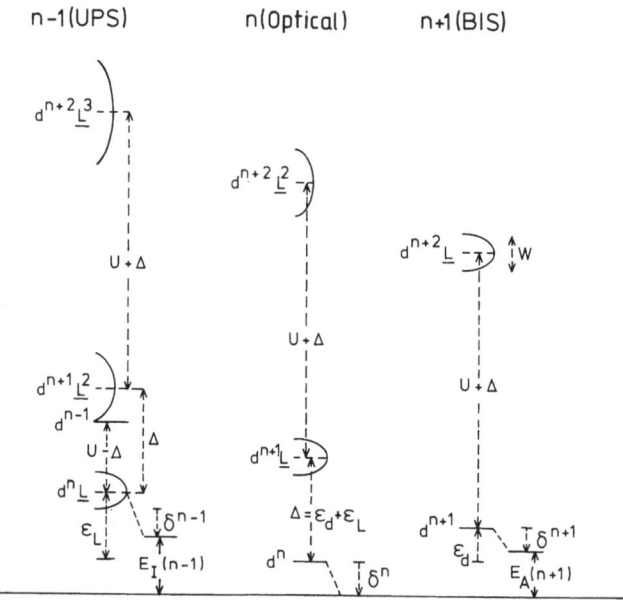

Figure 4. Total energy diagram indicating the states and continua entering the model calculation as described in the text. Hybridization shifts are also indicated.

It is not difficult to show that the superexchange interaction as calculated using a 3 center model is given by $2b^2(1/U_{ex} + 1/\Delta_{ex})$ rather than $2b^2/U$ as used by Anderson[5]. Here U_{ex} and Δ_{ex} are somewhat different from U and Δ using the Hund's rule ground state[8]. The resulting estimates of the Neel temperatures of the transition metal monoxides are compared with experiment and the large Δ approximation in Fig. 5.

There are however several problems here which should be adressed. We find that the parameters required to explain experimental data for the Ni dihalides depend strongly on the type of experiment. Δ is found to be considerable smaller in core level XPS[9] than in XAS or in optical spectroscopy[8]. Also a substantial variation in U and the hybridization interac-

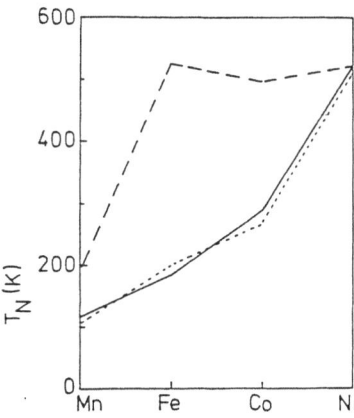

Figure 5. Experimental Néel temperatures of the monoxides (heavy line) compared to Anderson's results (dashed line) and our results (dotted line). J. Zaanen and G.A. Sawatzky, Can. J. of Phys. (in press) (1987).

Table 1. Estimates of the effective d-d Coulomb interaction (U) and the charge transfer energy (Δ) for the transition metal monoxides in the NaCl structure.

	U	Δ
Cu	5.1	4.0
Ni	7.3	6.0
Co	4.9	5.4
Fe	3.5	6.1
Mn	7.8	8.9

tion is found. These facts indicate that the model Hamiltonians we are using to describe core level experiments (especially XPS) probably omit one or more important interactions.

Based on NiO data and atomic spectroscopy we can make estimates of U and Δ for the other late TM divalent oxides. These are listed in table 1. For CuO the estimate is based on a NaCl structure rather than the real CuO structure. We can also estimate the parameters for systems like La_2CuO_4. Since the sum of the Madelung potentials on the Cu and O site is considerably larger[10] than used in the above NaCl structure Δ will be larger by about 2 eV. We then arrive at $U \simeq 5$ eV, $\Delta \simeq 6$ eV and a band gap of approximately 3 eV[11]. Interesting is that with the large hybridization of the $d_{x^2-y^2}$ orbitals because of the short CuO bond lengths, the lowest electron removal state would, like NiO, be a strongly mixed $Cu(d^8)$, $Cu(d^9)\underline{L}$ state. A good estimate of the parameters will however have to wait for detailed photoemission - inverse photoemission data.

There are a number of questions I would like to raise.

1. How far can we push the Anderson impurity model for transition metal compounds? We already have problems with the metallic systems like NiS.

2. Are the parameters obtained from core level spectroscopies transposable to band gap and optical spectroscopy descriptions?

3. In general we find for both the insulating and metallic systems that the monopole part of the d-d and f-f Coulomb interactions is strongly screened but the parts describing the spin and angular dependence are not. In view of this it is possible to get a negative U for triplet like pairing. Is this physically acceptable?

4. The paramters used in the Anderson like Hamiltonians are "highly effective" in that apparently large interactions (i.e. those that screen U from 20 to 5 eV) are not included explicitly in the Hamiltonian. Is this procedure justified when comparing low energy scale and high energy scale properties?

5. Although a treatment of the Anderson lattice problem for metals seems to be out of our reach at the moment perhaps a treatment of an Anderson lattice of impurities in a semiconductor is more tractable? If so, one could study the behaviour as the gap closes. For example, do the Anderson Holdane multiply charged impurity states in the gap of a semi conductor continuously develop into the Kondo resonance as the gap closes?

REFERENCES

1. See for example A.H. Wilson, Proc. Roy. Soc. A133:458 (1931).
    H.J. de Boer and E.J.W. Verwey, Proc. Phys. Soc. A49:59 (1937);
    N.F. Mott, Proc. Phys. Soc. Sect. A62:416 (1949);
    T. Oguchi, K. Terakura and A.R. Williams, Phys. Rev. B28:6443 (1983);
    A. Fujimori, E. Minami and S. Sugano, Phys. Rev. B29:5225 (1984);
    G.A. Sawatzky and J.W. Allen, Phys. Rev. Lett. 53:2239 (1984).

2. The band gap in late transition metal halides seems to scale with the electronegativity of the anion. e.g. see Ni dihalides
    C.R. Rondo, G.J. Arends and C. Haas, Phys. Rev. (in press) (1987).

3. CuS is a "p" type metal (superconductor)
    J.C.W. Folmer and F. Jellinek, J. Less Common Metals 76:153 (1980).

4. Excitonic $d^n \to d^{n*}$ or interband.
    S. Sugano, T. Tanabe and H. Kamimura in "Multiplets of transition metal ions in crystals (Academic, New York 1970).

5. O.W. Anderson, Solid State Physics 14:99 (1963).

6. J. Zaanen, G.A. Sawatzky and J.W. Allen, Phys. Rev. Lett. 55:418 (1985).

7. O. Gunnarsson and K. Schönhammer, Phys. Rev. B28:4315 (1983); 4815 (1985).

8. J. Zaanen and G.A. Sawatzky, Can. J. of Phys. (in press) 1987.

9. J. Zaanen, C. Westra and G.A. Sawatzky, Phys. Rev. B33:8060 (1986).
    G. van der Laan, J. Zaanen, G.A. Sawatzky, R. Karnatak and J.M. Esteva, Phys. Rev. B33:4253 (1986).

10. R.A. de Groot, private communication.

11. The band gap is smaller than in NiO because the d-p hybridization is larger.

NARROW BAND AND LOCALIZATION

Börje Johansson

Department of Physics, University of Uppsala
Box 530, S-751 21 UPPSALA, Sweden

Clearly the concept of energy bands is one of the most fundamental ingredients in the theory of condensed matter. It is now well established that energy band calculations are very useful in order to develop a microscopic understanding of materials. During the last 10-15 years there has been a very strong development of both computational techniques and computers so that now even bulk properties can be calculated from just a knowledge of the atomic constitution of the material (at least for moderately complex systems). The crystal structures of the elemental metals have also been successfully treated by means of energy band calculations. However, when we proceed to materials possessing so-called narrow bands the situation is less clear as regards the appropriateness of band structure calculations. When we go even further, namely to systems containing localized atomic-like configurations, then the conventional band theory approach fails to provide a proper description of the electronic properties. A well-known example of this is the CoO compound, which is a very good insulator, although normal band theory would predict it to be a metal. Thus it appears that the d-electrons in this system do not form extended Bloch states but rather are localized. This problem about localization versus itineracy was recognized long ago and especially Mott has stressed the difficulties involved. Therefore the transition from an insulator of this type to a metal (say under pressure) is nowadays commonly referred to as a Mott transition. However, this kind of insulator-metal phase transformation concept might be generalized to include also transitions in materials where the Mott delocalization of a particular type of electron occurs in the presence of other valence electrons. Such a phase change has been proposed to take place in cerium metal under pressure, where the delocalization process then would involve electrons in the 4f shell (the γ-α transition).

Experimentally there is now strong evidence that there is a fundamental change of the 5f electron behaviour as one proceeds through the series of actinide metals. Thus for the earlier actinide elements the 5f eelectrons have itinerant (band) properties, while for the heavier elements typical rare-earth-like behaviour is encountered (with localized $5f^n$) configurations). This localization takes place between plutonium and americium. It is significant that this change is accompanied by a large change of the atomic volume. Thus in plutonium the 5f electrons contribute with a metallic bonding, which gives rise to a low volume, while this bonding is absent in americium and therefore its equilibrium volume is comparatively large. Applications of pressure on the heavier actinide metals have produced crystallographic phase changes accompanied by large volume collapses. The change of 5f behaviour between plutonium and americium has been quite well accounted for by energy band calculations, where the possibility of a spin-polarization of the 5f electrons has been allowed for. For plutonium no spin-polarization is found in the calculations, while in americium there is a complete spin-polarization of the 5f electrons. Thus an almost filled spin up band and an empty spin down band is obtained. Since a filled band of this type gives no bonding contribution, the large volume for americium becomes well described and the spin-polarized solution mimics in a quite reasonable way the localized 5f properties.

A similar behaviour as for the actinides is encountered for the 3d monoxide compounds. Here, again as a function of the (3d) atomic number, a localization of the 3d electrons takes place. Thus for TiO and VO the 3d electrons are metallic, while in MnO and onwards local 3d properties are found. Also here there is a volume discontinuity between the lighter and heavier systems. Band calculations have been applied to these systems as well and again the change in behaviour is well accounted for by means of spin-polarized calculations. Also here the fully spin-polarized solution for MnO with five 3d-electrons gives essentially a filled majority band and an empty minority band and therefore there is no bonding contribution from the 3d electrons.

This in both the mentioned series of systems band calculations give a good account of the change of behaviour from itineracy to localization. One of the reasons for the success is that in both series the cross-over happens to take place near a half-filled band situation ($5f^7$ and $3d^5$, respectively). It is also true that if one proceeds to systems beyond the half-filled case (Bk or FeO), the calculations still give a spin-polarized solution as the stable state, but that now there is a bonding contribution from the open shell electrons (although considerably reduced compared to a non-polarized situation). Thus, in principle, it is only for the half-filled shell cases that the spin-polarized band calculations give a good description of the localization. As soon as we move away from the situation, spin polarization is not sufficient to account for the non-bonding property of the localized electrons. In the localized state the associated electronic configuration has, as already mentioned, atomic-like properties. By making a spin-polarized calculation one in fact essentially accounts for Hund's first rule for the atomic configuration. For a half-filled fully spin-polarized shell this is indeed a very good description of the atomic configuration f^7 and d^5. Since experimentally the atomic-like properties of the open shell are significant features of the localized configuration one has to account for Hund's second rule (i.e. by maximizing L under the constraint that S has already been maximized). Thus what is needed is a further generalization of the spin-polarized band calculations to include orbital polarization. This has so far not been done within a realistic band calculation, but appears to be necessary if one wants to describe localization in a more appropriate way for other cases than half-filled shell systems. A complete orbital polarization would thus be expected to give rise to filled bands and therefore a withdrawal of these bands from the Fermi energy. Thereby the insulating properties of these electrons in the localized phase on the low density side of a Mott transition will be reproduced. Furthermore the situation with filled bands will correspond to a non-bonding state. Thus the failure of spin polarization to properly account for non-bonding properties of localized states, except for the case of a half-filled shell, could thereby, it seems, be remedied. Hopefully this type of generalization of realistic band calculations will be made in the near future. Considering the usefulness of Stoner theory for the description of spin magnetism, it appears that a basic treatment of orbital polarization should also be most fruitful. In addition one could hope that the strong interest in the 3d oxides during the period 1960-1970 to some extent will be revived and that

new pressure experiments for the 3d monoxides will be performed in order
to identify and elucidate the expected Mott transitions. It is likely that
pressures of the order of 500 kbar to 1 Mbar will be necessary which, however, is within reach of the present techniques.

MODEL HAMILTONIANS AND HOW TO DETERMINE THEIR PARAMETERS

K. Schönhammer* and O. Gunnarsson+

* Institut für Theoretische Physik, Universitat Göttingen
3400 Göttingen, West Germany

+ Max-Planck-Institut für Festkörperforschung
7000 Stuttgart 80, West Germany

1. INTRODUCTION

It is straightforward to write down the Hamiltonian for the electrons
in molecules or solids within the Born-Oppenheimer approximation for the
nuclei. Unfortunately the resulting electronic many-body problem is notor-
iously hard to solve. The idea of mean field theory has therefore very
early been used in attempts to perform reasonable calculations for the
properties of atoms and molecules.[1] In quantum chemistry a Hartree-Fock
calculation is the usual starting point for configuration interaction (CI)
calculations, which can yield very accurate results for small molecules.
As this technique is not very well suited for solids, the Hohenberg-Kohn-
Sham density functional theory (DFT) represented a major step forward.[2] These
authors showed that the calculation of ground state properties can be formally exactly reduced to a problem of non-interacting electrons in an effective potential v_{eff}.

The great practical success of DFT is based on the fact that the local
density approximation (LDA) to v_{eff} works surprisingly well.[3] The self-
consistent solution of the Schrödinger equation for electrons in the potential v_{eff} yields as a by-product the one-electron eigenvalues ε_i. Although
these eigenvalues have no obvious meaning in the formalism, they are often
interpreted as excitation energies. This approach has been quite successful,
although it gives, for instance, a too small band-gap in semiconductors.

Quite generally the meaning of the density functional eigenvalues
becomes obscure, when electronic correlation effects are important. This,
is e.g., the case for the lanthanides. They have a partly filled atomic
4f-shell that is located mainly inside the 5s, 5p core. Therefore the Coulomb
interaction between two 4f electrons is large and a proper theoretical
treatment of this interaction is essential. For Ce compounds the f-occupancy
n_f in the ground state is close to one. Addition of a second f-electron in
Bremsstrahlung isochromat spectroscopy (BIS) leads experimentally to an
f^2-peak, which theoretically has no analogue in the density functional
eigenvalue spectrum. It is therefore necessary to perform a calculation of
one-electron Green's functions to produce the experimental spectra. Due to
the large f-f Coulomb energy it is unfortunately not sufficient to calculate
the electronic self-energy in perturbation theory or to use approximations
like the "GW"-approximation,[4] which seem to yield quite accurate results

for e.g. semiconductors. Because of these difficulties in treating the important f-f correlation with sufficient accuracy, it has become popular to abandon the exact Hamiltonian for the system and to introduce a simplified model Hamiltonian containing those interactions which are believed to be essential for the description of the system. For the stand-point of theoretical ("mathematical") physics such a model usually poses an interesting problem in its own right which can be studied in the whole parameter range to explore the physics contained in it. If one intends to use the model in comparison with experiments the question arises how to determine the parameters in the model. Different approaches to this question are the main topic of this paper. As an example we use the (generalized) Anderson impurity model for mixed valence systems.

2. THE MODEL AND ITS PARAMETERS

To describe the phenomenon of intermediate valence, a microscopic model must allow for the delocalization of the f-electrons by hopping to conduction electron states. The simplest model which contains this interplay between delocalization and strong f-f correlation is the Anderson impurity model[5] where one considers the f-level on <u>one</u> atom and its interaction with the conduction states.

$$H = \sum_{\vec{k},\alpha,\sigma} \varepsilon_{\vec{k},\alpha} n_{\vec{k},\alpha,\sigma} + \varepsilon_f \sum_{m,\sigma} n_{m\sigma} + \sum_{\vec{k},\alpha,\sigma,m} V_{\vec{k}\alpha,m} \psi_{m\sigma} \psi_{\vec{k}\alpha\sigma} + h.c.$$

$$+ \frac{U}{2} \sum_{(m,\sigma)\neq(m'\sigma')} n_{m\sigma} n_{m'\sigma'} \tag{1}$$

Here $\varepsilon_{\vec{k}\alpha}$ refers to a conduction state with wave vector k and band index α. The hybridization of the f-level with these conduction bands is described by hopping matrix elements $V_{\vec{k}\alpha m}$ where m is the orbital index. The "bare" energy of the 4f-level is ε_f and the last term describes the Coulomb interaction U between two 4f-electrons. Among the terms neglected is the Coulomb interaction between conduction states as they are rather extended and the Coulomb integrals are "not very large".

Can the parameters defined above be calculated using the knowledge of the "original" full microscopic Hamiltonian $H_{tot} = T + V_N + v_{ee}$? An attempt to bring H_{tot} into "model form" would start by defining a complete set $\{|i\rangle\}$ or orthonormal one-electron states which include the states $|k\alpha\rangle$ and $|f,m\rangle$. In second quantization, H_{tot} then contains operators like those in the model Hamiltonian (1), with <u>microscopic</u> expressions for the matrix elements, but many additional terms that are not small and certainly cannot just be neglected. Any attempt to get rid of them requires a renormalization of the matrix elements of the remaining terms. Furthermore, even if the space of one-electron states $\{|i\rangle\}$ is reduced to $|k,\alpha\rangle$ and $|f,m\rangle$ by some kind of projection technique, there will certainly remain other operators not included in (1), e.g. four-fermion operators such as operators describing the f-d Coulomb interaction U_{fd}. The neglect of these terms must again require a renormalization of matrix elements but in addition <u>dynamical</u> effects like those associated with f-d screening are no longer contained in the model. Such a simplification of the model is therefore only useful, if one considers a parameter range where the <u>dynamical</u> aspects of the f-d screening is unimportant. The obvious way to test the validity of the simpler model is to calculate physical quantities like spectral functions with the full model and then test if the simplified model with renormalized parameters leads to similar results. Unfortunately the model including the f-d Coulomb interaction is much more difficult to solve with a reasonable accuracy and therefore the above procedure cannot be carried out in this case. This shows a serious problem involved in a microscopic determination of parameters in a model Hamiltonian: It seems to require a proper solution of the problem with a Hamiltonian containing more complicated interactions. But to simplify the solution for a given Hamiltonian was the main reason to introduce a simplified model Hamiltonian. This seems to indicate that it is more useful to consider a model Hamiltonian like (2) as a 'phenomenological' Hamiltonian with parameters to be determined by some experiments and then to <u>predict</u> other measurable quantities. This point of view is discussed in Section 4. In the next section we first show how the elimination of interaction terms can be approximately carried out explicitly if the f-d screening is treated theoretically in a plasmon picture and the plasmon is described as a single boson. This example shows ('a posteriori') that the renormalization of parameters with the neglect of the "f-d"-interaction does not really require the 'dynamical' solution of the full model, but can be obtained from simple 'static' considerations.

3. COMPETITION BETWEEN SCREENING MECHANISMS

In this section we focus on the calculation of photoelectron spectra, especially the emission from the f-states. As the long range Coulomb-interaction is not properly contained in the model (2) one has to discuss how the screening of the f-hole can be described. Two screening mechanisms have to be distinguished: Screening by conduction electrons and screening by "filling the f-hole".[6] A somewhat similar situation with two screening mechanisms can occur in core-hole spectroscopy of adsorbates,[7] where the screening can either be "image screening" or screening by charge transfer to the affinity level of the adsorbate which is unoccupied in the ground state.[8] Describing the "image screening" due to surface plasmons by a single boson mode, we have calculated the core spectral function with <u>both</u> screening mechanisms as well as with the charge transfer mechanism only.[8]

The results are shown in Fig. 1. The dotted curve shows the result for the charge transfer screening (CTS) only. The two peaks have been called "poorly screened" and "well screened" peaks. Switching on the additional plasmon mechanism has essentially two effects: A shift of the former "poorly screened" peak by $e^2/4d$, and a plasmon satellite as an additional "dynamic"

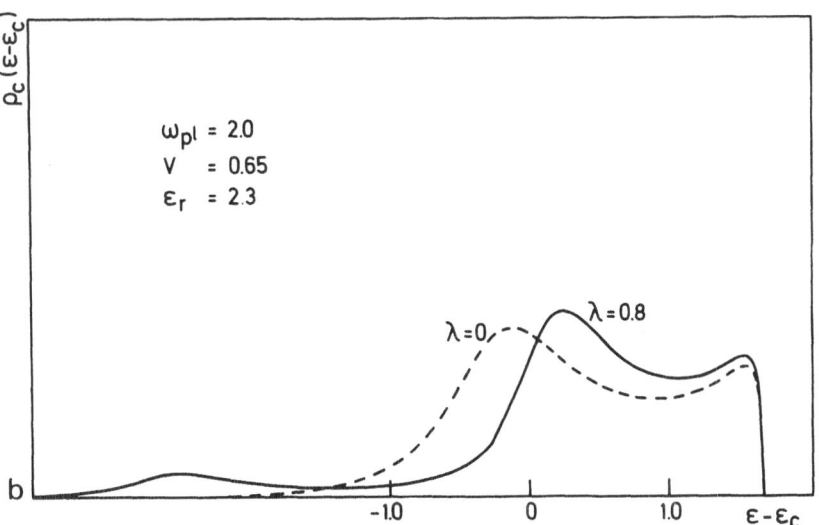

Fig. 1.   The core hole spectral function for two values of the plasmons coupling strength λ. The plasmon energy is $\omega pl = 2$.

feature in the spectrum. For the parameters used in Fig. 1 the satellite is not very strong. If experimentally one does not probe this "high energy region", a very similar spectrum as the full curve can be obtained with the CTS only but with a renormalization of the core level energy $\varepsilon_c \rightarrow \varepsilon_c + e^2/4d$ and a renormalization of the screening orbital energy $\varepsilon_a \rightarrow \varepsilon_a + e^2/4d$. The shift $e^2/4d(=\lambda^2/\omega_{pl}$ in the boson model) is just the <u>static</u> relaxation energy.

In the lanthanides the f-level plays the role of the screening orbital. In Ce compounds the f-occupancy n_f in the ground state is of order one and in addition to core level spectra one can study photoemission from the f-level itself. An analytical discussion of the quality of the dynamical elimination of the conduction electron screening can be obtained for the f-hole spectral function when the large degeneracy N_f of the f-level is taken into account. If the conduction screening is again described by a single boson $\omega_{pl}b^+b$, the screening interaction is of the type

$$\lambda(b+b^+)(n_f - \sum_\nu n_\nu).$$

In the limit $N_f \rightarrow \infty$, $N_f\Delta \rightarrow$ const.,

$$\Delta(\varepsilon) = \sum_{\vec{k},\alpha} |V_{\vec{k}\alpha}|^2 \delta(\varepsilon - \varepsilon_k)$$

the f-spectral function can be calculated exactly.[9] It turns out that if the plasmon frequency is large enough, the f-spectrum (except for the plasmon satellite region) takes the form of the spectrum without the plasmon mechanism but with renormalized parameters: $\varepsilon_f \rightarrow \varepsilon_f + (2n_f-1) \lambda^2/\omega_{pl} \simeq \varepsilon_f + \lambda^2/\omega_{pl}$, and a weak renormalization of the hopping matrix elements.[9]

The Anderson model (2) in which screening by conduction electrons is only indirectly included via the renormalizations discussed above, leads to an f-photoemission spectrum which can have a pronounced <u>two peak structure</u>[10] as found experimentally in many Ce compounds: A "Lorentzian peak" close to ε_f, which is a "conduction electron screened peak" and a peak close to the Fermi energy which is the onset of the Kondo peak seen in BIS "above the Fermi energy". The final states corresponding to this peak are of the type $\psi_{\vec{k}_F\alpha} |\phi_0\rangle$. Near the impurity these states look like the ground state, i.e. the f-hole is screened by "filling it again".[10]

The opposite point of view concerning the importance of the two screening mechanism has been taken by Liu and Ho[11] and by Riseborough[12], who emphasized the dynamic aspect of the conduction electron screening. They neglect the f-d hybridization, i.e. they eliminate the possibility of f-screening. In their interpretation the two peaks in the f-spectrum are "poorly" and "well" screened with respect to the d screening mechanism. We have, however, shown that the "poorly screened" peak has little weight (the dynamical effects are small) if a realistic width and filling of the d-band are used and if the value of U_{fd} is chosen such that the experimental peak separation is reproduced.[13]

A conclusion of this section is that it is not always necessary to fully solve the "complicated model" to determine renormalizations. If the dynamical effects of the mechanism which one wants to neglect are unimportant, the proper renormalization can be obtained for simpler static considerations. In this way Herbst et al.[14] have calculated the renormalization of U and ε_f due to "U_{fd}-screening" and other interactions neglected in (1). The renormalization of hopping matrix elements is less obvious. It

130

however appears that a conclusion of these quantities in the LDA, without any further renormalization, give surprisingly good results.[15]

4. THE PHENOMENOLOGICAL POINT OF VIEW

Above we discussed how model Hamiltonians can be obtained from the full microscopic Hamiltonian by elimination of degrees of freedom. An alternative approach is to guess the important degrees of freedom and the form of the most important interactions. Considerations concerning the feasibility of a proper solution may also play a role. This is probably the way Anderson arrived at his model. For an experimentalist such a model Hamiltonian is certainly useful if it has 'predictive power'. If the model does not contain too many parameters, it may be possible to completely determine these parameters by comparison of a few calculated properties with the measured ones and then to predict other physical quantities for the same system.

We have adopted this point of view quite successfully to understand the electronic properties of mixed valence Ce compounds.[10] For details we refer to recent review articles.[13,16] It is quite easy to determine ε_f, U and the average hybridization strength by comparison with experimental core level spectra. A more delicate issue is the determination of the function $\Delta(\varepsilon) \approx \Delta_{av} \rho_{Band}(\varepsilon)$. In the purely phenomenological point of view the function $\rho_{Band}(\varepsilon)$ is determined again by comparison with experiment, i.e. valence photoemission in a photon energy range where f-emission is negligible. Unfortunately this procedure involves simplifying assumptions about dipole- and hopping matrix elements.

It is therefore interesting to combine the phenomenological point of view with "ab initio" calculations, e.g. DFT eigenvalues

$$\varepsilon^{DFT}_{\vec{k},\alpha}$$

as the conduction band energies.[15] There has been some promising work done along these lines, although there are unresolved theoretical questions, like how to avoid overcounting etc.

Let us finally mention another important aspect of model Hamiltonians like (1). They can lead to simple qualitative pictures of the physics of the systems studied. An example is the emergence of a low energy scale in (2) in connection with the Kondo resonance which allows us to reconcile experimental findings in spectroscopy and in thermodynamic measurements. This may help us to understand the physics of these systems, whatever that means.

REFERENCES

1.   D.R. Hartree, Proc. Camb. Phil. Soc. 24:89 (1928).
2.   P. Hohenberg and W. Kohn, Phys. Rev. 136:B864 (1964);
     W. Kohn and L.J. Sham, Phys. Rev. 140:A1133 (1965).
3.   U. von Barth and A.R. Williams, in "Theory of the Inhomogeneous Electron Gas", Eds. S. Lundqvist and N.H. March (Plenum, New York, 1983).
4.   L. Hedin, Phys. Rev. 139:A796 (1965).
5.   P.W. Anderson, Phys. Rev. 124:41 (1961).
6.   A. Kotani and Y. Toyozawa, J. Phys. Soc. Japan 37:563 (1974).
7.   K. Schönhammer and O. Gunnarsson, Solid State Commun. 23:691 (1977);
     26:399 (1978).
8.   O. Gunnarsson and K. Schönhammer, Solid State Commun. 26:147 (1978);

K. Schönhammer and O. Gunnarsson, $\underline{Z}$. $\underline{Phys}$. B30:297 (1987).

9. K. Schönhammer and O. Gunnarsson, $\underline{Phys}$. $\underline{Rev}$. B30:3141 (1984).

10. O. Gunnarsson and K. Schönhammer, $\underline{Phys}$. $\underline{Rev}$. $\underline{Lett}$. 50:604 (1963); $\underline{Phys}$. $\underline{Rev}$. B28:4315 (1983); B31:4815 (1985).

11. S.H. Liu and K.-M. Ho, $\underline{Phys}$. $\underline{Rev}$. B26:7082 (1982); B28:4220 (1983).

12. P.S. Riseborough, $\underline{J}$. $\underline{Magn}$. $\underline{Magn}$. $\underline{Mater}$. 47-48:271 (1985).

13. J.W. Allen, S.J. Oh, O. Gunnarsson, K. Schönhammer, M.B. Maple, M.S. Torikachvili and I. Lindau, $\underline{Adv}$. $\underline{Phys}$. 35:275 (1986).

14. J.F. Herbst, R.E. Watson, and J.W. Wilkins, $\underline{Phys}$. $\underline{Rev}$. B17:3089 (1978); J.F. Herbst and J.W. Wilkins, $\underline{Phys}$. $\underline{Rev}$. $\underline{Lett}$. 43:1760 (1979).

15. O. Sakai, H. Takahashi, M. Takeshige, and T. Kasuya, $\underline{Solid}$ $\underline{State}$ $\underline{Commun}$. 52:997 (1984).

    R. Monnier, L. Degiorgi, and D.D. Koelling, $\underline{Phys}$. $\underline{Rev}$. $\underline{Lett}$. 56:2744 (1986).

    L. Degiorgi, T. Greber, F. Hulliger, R. Monnier, L. Schlapbach, and B.T. Thole, $\underline{Europhysics}$ $\underline{Letter}$.

16. O. Gunnarsson and K. Schönhammer, Handbook on the Physics and Chemistry of Rare Earths (eds. K. Gschneider, L. Eyring and S. Hüfner), Vol. 10, North-Holland, Amsterdam (1986).

LOCAL DENSITY CALCULATED PARAMETERS FOR THE ANDERSON HAMILTONIAN

A.K. McMahan* and R.M. Martin[†]

* University of California, Lawrence Livermore National
Laboratory, Livermore, CA. 94550, USA

[†] Xerox Palo Alto Research Center, 333 Coyote Hill Road
Palo Alto, CA 94304, USA

Local density functional theory has recently been used to provide
first principles calculation of parameters entering the impurity Anderson
Hamiltonian, with promising results. Examples are presented here for the
case of the rare earth dioxides CeO_2 and PrO_2.

The Anderson impurity Hamiltonian has shown considerable success in
accounting for spectroscopic[1] and low energy electronic properties[2-4] of
rare earth elements and compounds and some insulating transition metal
compounds.[5-12] It is believed that the lattice version of the Hamiltonian
may elucidate the behaviour of heavy Fermion materials[13]. The key features
incorporated in the model are the localized state with energy ε_f, the Coulomb
interaction U, and the hybridization $V(\varepsilon)$ with a continuum of other delocalized states. Recently, there have emerged new methods to solve this many-
body Hamiltonian, particularly from the work of Gunnarsson and Schönhammer[14].
By choosing the parameters entering the impurity Hamiltonian empirically,
often by fitting spectroscopic data, it has been possible to correlate
very different properties that depend upon the many-body effects. These
advances have led to improved understanding of the behaviour of realistic
systems.

The subject of the present paper is the calculation of the parameters
in the Anderson Hamiltonian from theory, together with predicted properties
of selected systems. The problems are to identify the appropriate localized
state, calculate its energy for different integral occupations, and determine
the hybridization with other states. We apply the methods to CeO_2 and PrO_2
as examples of the problems in insulators with non-magnetic and magnetic
solutions.

Density functional theory should in principle give the correct ground
state energy, including the total energies for a localized state at a single
site in which an electron has been added or removed. From differences between
such total energies one can obtain ε_f and U. Similar calculations of total
energy differences are performed for atoms.[15] However, in a solid one has
the important additional effects of including the screening of all the
other electrons in the values of the parameters appropriate to use in the
Anderson Hamiltonian. Although the Hamiltonian does not explicitly include
the Coulomb interactions between the localized and the band-like electrons,

the point is that they are included in the effective parameters. Note also that these interactions also play crucial roles in other many-body problems, such as core spectra and models proposed for Ce by Falicov and coworkers[16], Liu and Ho[17], and others[18]. We do not have space to deal with these here.

Some caution must be exercised in using the local approximation to density functional theory. Nevertheless, recent local calculations of the parameters ε_f and U have compared favorably to features in spectroscopic data.[19-22] The computational problem of dealing with a single impurity in an infinite system is usually handled by supercell[23] calculations in which a lattice of differing sizes are extrapolated to infinite impurity separation. The key issue is the accuracy of such a method in including all the screening effects of the other electrons. In a metal there is complete screening on a length scale of the order a Thomas-Fermi length. Thus the localized state in the Anderson Hamiltonian is actually a screened state with no long range interactions. In fact, Min et al.[22] showed that the supercell calculations for elemental Ce converge for very small cells. For insulators, on the other hand, there is a long range Coulomb interaction with is screened by the dielectric function.[24] For these cases more care is needed to correctly reach the asymptotic limit. Our work to establish the correct limits for representative cases will be described elsewhere in some detail. It might also be noted that the localized states used for electron removal and addition are generally treated in a manner similar to core states: They are determined self-consistently from the one-electron potential, but are prohibited from hybridizing with other states, so that the calculated U is the desired bare ($V(\varepsilon)=0$) Coulomb interaction. It is, of course, still screened by polarization of the delocalized electrons.

The hybridization $V(\varepsilon)$ results from overlap of electron orbitals on different sites and in general of different angular momenta. This is precisely the role played by hybridization in determining crystalline Bloch states in standard one-electron theory. Thus one may extract the hybridization matrix elements of interest, for example, by a tight-binding fit to the local density one-electron bands, as has recently been done for YbP.[25] There is a difficulty, however, because the matrix element $V(\varepsilon)$ in the Anderson Hamiltonian may depend upon the Coulomb interaction U. Hirst, for example, has given cases for Sm and Eu in which the matrix element is qualitatively reduced by correlations of many f electrons on the localized site.[26] These effects are expected to be much less important in cases in which the 4f level is nearly full or nearly empty, as in Ce and Pr. At this stage, therefore, we consider the hybridization as found in local density theory, which includes the Coulomb effects only in an average sense. Note also that $V(\varepsilon)$ is not unique, as it depends on the manner in which the impurity and host states are decoupled. This is not a problem in practice, however, as there are sensible criteria, such as choosing basis states for which the impurity couples only to relatively near host neighbors.

In spite of these caveats, first principles calculation of ε_f, U, and $V(\varepsilon)$ by local density functional theory can be of considerable value, as we illustrate now for the case of the rare earth dioxides CeO_2 and PrO_2. These are fluorite structure insulators with filled oxygen 2p valence bands. As there is only small direct overlap between 4f orbitals on different rare earth (RE) sites, these atoms may be considered independent impurities which hybridize with the host O-2p bands. The impurity Anderson Hamiltonian is thus an appropriate model for the overall features, bearing in mind that the translational invariance of the lattice is important for some properties. previous work, mostly on CeO_2, includes ordinary local density band calculation of the density of states,[27] and empirical solutions[28-31] of the Anderson Hamiltonian in order to fit spectroscopic data.[31-33] We have obtained[34] all parameters in the present work using the linear muffin-tin orbitals[35,36] implementation of local density theory, and then solved

Table 1. Results for CeO_2 and PrO_2. Position of the 4f level ε_f and the Coulomb interaction U were obtained from LMTO supercell calculations. Second order solution of the impurity Hamiltonian yields ground state 4f occupation $\langle n_f \rangle$, and a lowest excitation energy $E(\Gamma_7)-E(\Gamma_8)$ for PrO_2. Calculated $f^n \rightarrow f^{n+1}$ inverse photoemission peaks are composed of five lines ranging from $\varepsilon(\Gamma_8)$ to $\varepsilon(\Gamma'_7)$. Energies ε, U, and E are in eV. Electron addition energies ε are relative to the top of the 0-2p band. Values in parenthesis from Ref. 31 are provided for comparison in the case of CeO_2.

	n	$\varepsilon_f + nU$	U	$\langle n_f \rangle$	$E(\Gamma_7)-E(\Gamma_8)$	$\varepsilon(\Gamma_8)$	$\varepsilon(\Gamma'_7)$
CeO_2	0	1.5	9.9	0.48		3.52	4.00
		(0.5)	(7.0)	(≈0.5)			
PrO_2	1	0.3	10.6	1.54	0.157	2.33	2.84

the Anderson Hamiltonian to second order following the approach of Gunnarsson and Schönhammer.[14]

The results for the 4f energy (relative to the top of the 0-2p band) and U are given in Table I. Note that for Ce the f state is above the 0 band; therefore, it there were no hybridization, it would be empty and CeO_2 would properly be termed "tetravalent". For Pr the first 4f electron is bound by nearly 11 eV and it is more relevant to consider the energy to add the second f electron $\varepsilon_f + U$. In our calculations this is found also to be above, but closer to the top of the 0-2p band. We should point out that the differences between PrO_2 and CeO_2 should be more accurate than the absolute energies of the f states relative to the 0-2p states. Our present uncertainty in ε_f for CeO_2 arising from supercell convergence, for example, is about ±1 eV, while that in the 1.2 eV difference between ε_f for CeO_2 and $\varepsilon_f + U$ for PrO_2 is only about 0.1 eV. In addition to screening in the supercell, the local density approximation may affect energy differences between very different states like the 4f and 0-2p. We expect these effects also to cancel between PrO_2 and CeO_2. Thus, we think the most definitive prediction of these calculations is that the gap in CeO_2 should be about 1 eV larger than in PrO_2. Although we know of no accurate experimental measurements of the gaps, our results are in reasonable accord with the fact that CeO_2 is described as pale yellow; PrO_2; as black.[37] We can also compare at this point with the fitted parameters of Ref. 31 for CeO_2, which are given in parentheses in Table I. The values of U are both sufficiently large (7 vs. 10) to lead to qualitatively similar conclusions. On a crude scale ε_f is also close (0.5 vs. 1.5); however, this difference leads to an important 1 eV increase in the insulating gap. Potentially our result is more nearly correct since the final gap reported in Ref. 31 is significantly smaller than the experimental gap in CeO_2.

The effect of the hybridization is very large as seen in Table I both from 4f occupancy $\langle n_f \rangle$ and by the opening of the gap from ε_f ($\varepsilon_f + U$ for PrO_2) to the value $\varepsilon(\Gamma_8)$, the lowest level in the inverse photoemission spectrum. One conclusion is that the properties of these materials are governed by the combination of Coulomb and hybridization effects, both of which are large. Perhaps the most striking result to mention here from our local density calculations is structure found in the hybridization $V_\mu(\varepsilon)$, generally taken in empirical modelling to be independent of both position ε in the 0-2p band, and of the symmetry μ of the crystal field split 4f

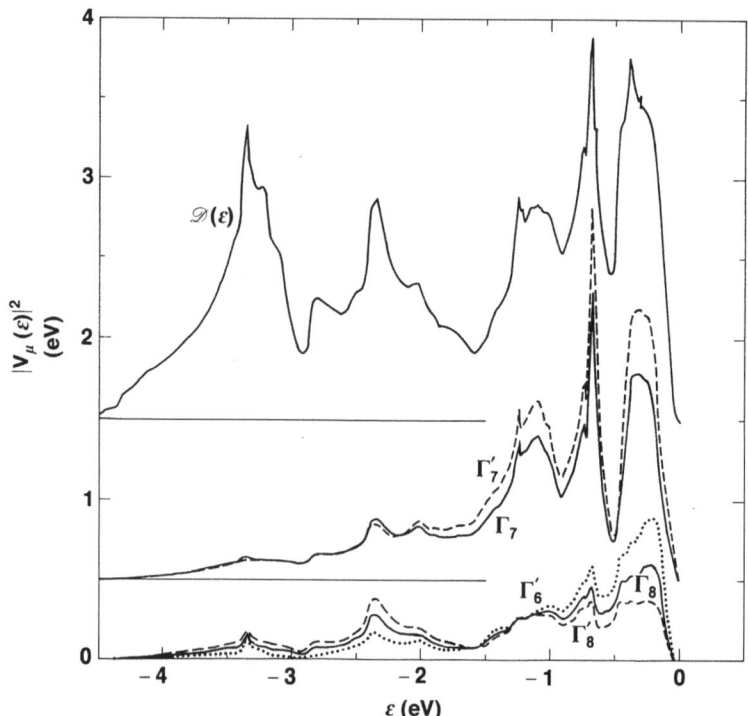

Fig. 1.   Calculated Ce-4f to O-2p hybridization $|V_\mu(\varepsilon)|^2$, and O-2p density of states $D(\varepsilon)$ in CeO_2, versus energy ε relative to the top of the O-2p band. Relativistic symmetry labels μ corresponding to $j+5/2$ ($j=7/2$) states are indicated without (with) primes. For visual clarity, the Γ_7 curves have been shifted upward 0.5 states per eV. $D(\varepsilon)$ (arbitrary units) has also been shifted upward.

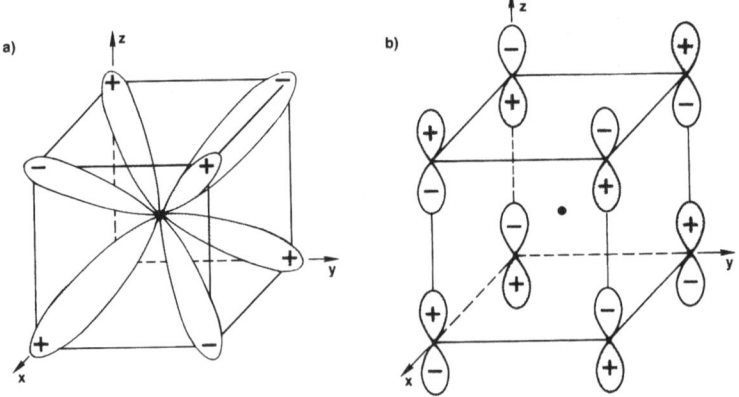

Fig. 2.   Sketches of a) RE-4f and b)O-2p orbitals in the fluorite structure. The former is odd under reflection through the xy plane passing through the body center, and thus couples strongly to the similarly odd O configuration in b), which represents antibonding chains of O atoms in the z direction. The sketch in a) corresponds to the nonrelativistic ($\Gamma'_2 = A_{2u}$) limit of Γ_7 (properly written Γ^-_7); that in b), to the nonrelativistic X'_2 state at the top of the O-2p band.

level $\varepsilon_{f\mu}$. As seen by comparison of $|V_\mu(\varepsilon)|^2$ to the O-2p density of states in Fig. 1, however, the RE level couples mainly to the top of the O-2p band. The reason is evident from the sketches of RE-4f and O-2p orbitals seen in Figs. 2a and 2b, respectively. Each RE atom lies at the body center of a cube of eight O's. As the RE-4f orbital is odd under reflection through, e.g., the xy plane passing through the body center, it couples strongly only with O-2p band states which are similarly odd under this reflection. As seen in Fig. 2b, such states are antibonding and will occur near the top of the O-2p band.

Dependence of the hybridization on the local symmetry μ of the 4f orbital leads, for example, to an $f^0 \to f^1$ inverse photoemission peak for CeO_2 composed of five discrete lines, and this provides a genuine theoretical prediction for the width of this peak. This is about 0.5 eV as seen from the spread between the $\varepsilon(\Gamma_8)$ and the $\varepsilon(\Gamma_7)$ lines. For PrO_2, on the other hand, there is a much more important effect. The low energy states are magnetic and the different magnitude of the hybridization for the possible symmetry states leads to the capability of calculating the magnetic PrO_2 ground state as well as the excitation energy to the low energy excited state measurable in neutron scattering. While the ground state wavefunction has the symmetry μ of an f^1 configuration $|f_\mu\rangle$, e.g., $\mu = \Gamma_7$ or Γ_8 for a j=5/2 f electron in a cubic environment, the hybridization $V_\nu(\varepsilon)$ admixes f^2 configurations $|f_\mu;f_\nu h_\nu\rangle$ in which an additional f electron and a compensating oxygen hole have symmetry ν. More of these admixed f^2 configurations are coupled by the larger Γ_7 hybridization (see Fig. 1) if $\mu = \Gamma_8$, which is then the predicted (and observed[38]) ground state. True crystal field effects (splittings in the bare $\varepsilon_{f\mu}$ due purely to the electrostatic potential at the Pr site) also favour Γ_8; however, their affect is considerably smaller in the present case. This sixth column of Table I gives the excitation energy $E(\Gamma_7)-E(\Gamma_8)$ for PrO_2 of 0.157 eV, which may be compared to 0.13 eV reported by Kern et al.[38] using neutron scattering. This number is not very sensitive to $\varepsilon_f + U$, but it is crucially dependent on V, scaling as V^2. The results for this sensitive test for PrO_2 leads us to conclude that our local density calculations of the hybridization, together with the Gunnarsson-Schönhammer method, are giving reliable results for both of these rare earth oxides.

In summary, first principles calculation of the parameters entering the impurity Anderson Hamiltonian can provide a richness of structure which is not easily achieved in empirical modelling of experimental data. This is especially true of the hybridization, although systematics deduced from calculated ε_f and U can also be valuable. We find for example, that $\varepsilon_f + U$ is closer to the top of the O-2p band in PrO_2, than is the counterpart level ε_f in the case of CeO_2, suggesting more pronounced many-body effects in the former material. PrO_2 would thus make a nice candidate for photoemission and inverse photoemission measurements. We believe that other systems, e.g., heavy Fermion materials should be reasonably described by such a combination of realistic density functional calculations of the basic energies and interactions, coupled with many-body solutions of the Anderson-type model hamiltonians.

This work has been supported by Lawrence Livermore National Laboratory under Contract W-7405-Eng-48 to the U.S. Department of Energy, and by the U.S. Office of Naval Research under Contract N00014-82-C-0244.

REFERENCES

1.   J.W. Allen, S.-J. Oh, O. Gunnarsson, K. Schönhammer, M.B. Maple, M.S. Torikachivili and I. Lindau, Adv. in Physics 35:275 91986).
2.   J.W. Allen and R.M. Martin, Phys. Rev. Lett. 49:1106 (1982).

3. M. Lavagna, C. Lacroix and M. Cyrot, Phys. Lett. 90A:210 (1982); J. Phys. F13:1007 (1983).

4. R.M. Martin and J.W. Allen, J. Magn. and Magn. Mater. 47-48:257 91985).

5. L.C. Davis, J. Appl. Phys. 59:525 (1986).

6. G. van der Laan, C. Westra, C. Haas and G.A. Sawatzky, Phys. Rev. B23:4369 (1981).

7. G. van der Laan, Solid State Commun. 42:165 (1982).

8. G.A. Sawatzky in: "Studies in Inorganic Chemistry", Vol. 3 (Elsevier, Amsterdam, 1983), p.3.

9. A. Fujimori and F. Minamai, Phys. Rev. B30:957 (1984).

10. G.A. Sawatzky and J.W. Allen, Phys. Rev. Lett. 53:2339 (1984).

11. J. Zaanen, G.A. Sawatzky and J.W. Allen, Phys. Rev. Lett. 55:418 (1985).

12. J. Zaanen, G.A. Sawatzky and J.W. Allen, J. Magn. and Mag. Mat. 54-57:607 (1986).

13. P.A. Lee, T.M. Rice, J.W. Serene, L.J. Sham and J.W. Wilkins, Comments Cond. Mat. Phys. 12:99 (1986).

14. O. Gunnarsson and K. Schönhammer, Phys. Rev. B28:4315 (1983).

15. R.O. Jones and O. Gunnarsson, Phys. Rev. Lett. 55:107 (1985).

16. R. Ramirez and L.M. Falicov, Phys. Rev. B3:2425 (1971); L.M. Falicov and J.C. Kimball, Phys. Rev. Lett. 22:997 (1967).

17. S.H. Liu and K.-M. Ho, Phys. Rev. B30:3039 (1984).

18. F.D.M. Haldane, Phys. Rev. B15:2477 (1977).

19. M.R. Norman, D.D. Koelling, A.J. Freeman, H.J.F. Jansen, B.I. Min, T. Oguchi, and L. Ye, Phys. Rev. Lett. 53: 1673 (1984).

20. M.R. Norman, D.D. Koelling and A.J. Freeman, Phys. Rev. B31:6251 (1985).

21. M.R. Norman, Phys. Rev. B31:6261 (1985).

22. B.I. Min, H.J.F. Jansen, T. Oguchi and A.J. Freeman, Phys. Rev. B33:8005 (1986).

23. A. Zunger and A.J. Freeman, Phys. Rev. B16:2901 (1977).

24. N.F. Mott and M.J. Littleton, Trans. Far. Soc. 34:485 (1938).

25. R. Monnier, L. Degiorgi and D.D. Koelling, Phys. Rev. Lett. 56:2744 (1986).

26. L.L. Hirst, Phys. Rev. B15:1 (1977).

27. D.D. Koelling, A.M. Boring and J.H. Wood, Solid State Commun.47:227 (1983).

28. A. Fujimori, Phys. Rev. B28:2281 (1983).

29. A. Kotani, H. Mizuta and T. Jo, Solid State Commun. 53:805 (1985).

30. T. Jo and A. Kotani, Solid State Commun. 54:451 (1985).

31. E. Wuilloud, B. Delley, W.-D. Schneider and Y. Baer, Phys. Rev. Lett. 53:202 (1984).

32. J.W. Allen, J. Magn. and Mag. Mater. 47-48:168 (1985).

33. T. Hanyu, T. Muyahara, T. Kamada, H. Ishii, M. Yanagihara, H. Kato, K. Naito, S. Suzuki and T. Ishii, J. Magn. and Mag. Mater. 52:193 (1985).

34. Scalar relativistic, total energy, supercell calculations were used to obtain the term average ε_f of the spin-orbit/crystal field split level $\varepsilon_{f\mu}$, and U. The splittings were determined by separate calculation. The hybridization was calculated using Eq.(2) of Ref. 25, with matrix elements V obtained by an approximation $H_0 + V$ to O.K. Andersen's second order Hamiltonian [Ref. 35], where H_0 excludes coupling between RE-4f and other states by removing such coupling from the structure constants, and V is nonzero only between RE-4f and other states. These results are sufficient to provide $|V_\mu(\varepsilon)|^2$ with fully relativistic symmetry index μ, given negligible spin-orbit splitting in the O-2p band.

35. O.K. Andersen, Phys. Rev. B12:3060 (1975); O.K. Andersen, O. Jepsen and D. Glötzel, in: "Highlights of Condensed-Matter Theory", LXXXIX Corso (Soc. Italiana di Fisica, Bologna, Italy, 1985).

36. H.L. Skriver, "The LMTO Method" (Springer, Berlin, 1984).

37. L. Eyring, in: "Handbook on the Physics and Chemistry of Rare Earths", edited by K.A. Gschneidner, Jr., and L. Eyring (North-Holland, Amsterdam, 1979), Chapter 27, Table 27.1

38. S. Kern, C.-K. Loong and G.H. Lander, Phys. Rev. B232:3051 (1985).

NOTES FROM GROUP DISCUSSIONS ON DENSITY FUNCTIONAL AND BANDSTRUCTURE CAL
CULATIONS FOR NARROW-BAND MATERIALS

These notes can of course not cover all questions that came up during the discussion. The major issues were as follows:

(i) Ground-state properties and density functional theory
    Successes and failures of the local-density approximation (LDA) to ground-state properties were discussed in some detail. The general view seemed to be that the LDA in general works quite well for such properties although we do not fully understand why. Disturbing exceptions such as the improper description of the α - γ transition in Ce and the vanishing gap in CoO were also discussed. It was argued that some failures could be related to errors in the eigenvalues, which would lead to an incorrect population of the density-functional orbitals.

(ii) Are there problems with the LDA Fermi surfaces?
    Also this question turned out to be rather uncontroversial. Thus, it was concluded that the LDA mostly gives quite reasonable Fermi surfaces, although there are of course exceptions, and several examples were quoted by de Groot and by others. It was argued that the Fermi surface is rather constrained by symmetry requirements and that this at least in part could explain the agreements. The striking agreement obtained for the heavy-fermion system UPt_3,[1] however, was not easily understood.

(iii) Why is the Hartree-Fock gap in insulators so large and the LDA gap so small?
    The general opinion seemed to be that the large HF gaps are a consequence of the unphysically long range of the corresponding selfenergy. It was more difficult to obtain simple arguments why local potentials always seem to decrease the separation between occupied and unoccupied states. A suggestion came up that the Talman scheme,[2] which in a sense gives the best local potential for describing exchange effects, could shed some light on this question.

(iv) Excitation energies from the selfenergy
    One of the issues here was the possible usefulness of the simple "GW" approximation[3] to quasiparticle excitations. There are no results as yet for narrow-band materials, but the proponents (mainly Almbladh) argued that at the successes in simpler insulators and semiconductors[4] give some hope that the GW scheme could be useful also for 3d systems. The opponents (mainly Peter Fulde) pointed out that the GW scheme grossly overestimates

the correlation energy in the bonds even in systems as simple as silicon because higher-order exchange diagrams are left out.[5] The apparent contradiction between these results[4,5] could not really be resolved, but it was generally agreed upon that the GW scheme would not be reliable when most of the physics lies in short-range correlations.

(v) Calculations of parameters in model Hamiltonians

Are physically sound ways to relate model parameters to quantities obtained from, say, LDA calculations? A common view seemed to be that there are several ways to extract model parameters in practice but that the theoretical justification is often weak. Thus, ΔSCF calculations of d or f electrons energies and Coulomb-U parameters have been found to yield useful results[6] but the f-electron is forced to be localized "by hand" in a way which is not unambiguous. Several alternative ways were also discussed, such as Williams and Kübler's idea to infer energy shifts of d levels in NiO from an Ni atom in a different host,[7] and Dederichs' density-functional treatment of constrained systems.[8] A nice way to deduce parameters for the Anderson impurity Hamiltonian including also the hybridization was presented by A.K. McMahan[9].

(vi) Magnetic "Mott" insulators NiO etc.

Most of the discussions here were centered on the question of the energy gap. One point of view (Brandow) was that the small LDA gaps would mean that the agreement obtained for ground-state properties is somewhat fortuitous. An opposite point of view (Almbladh) was that the rather good spin densities obtained from the LDA are difficult to reconcile with large errors of several eV in eigenvalue differences, and that also the "exact" density-funcitonal gap is probably much smaller than the correct one. As to the charge-transfer character of the gap in NiO, A.R. Williams maintained that the top of the valence band has mainly d character,[7] and G.A. Sawatzky maintained that it involves large p-to-d charge transfer.[10] However, no significant new theoretical or experimental results pertaining to this interesting question came up during the discussions.

Finally, Baird Brandow gave an overview of alternative approaches to the magnetic insulators.[11]

REFERENCES

1. G.S. Wang, M.R. Norman, R.C. Albers, A.M. Boring, W.E. Picket, H. Krakauer, and N.E. Christensen, Phys. Rev. B, to be published.
2. J.D. Talman and W.F. Chadwick, Phys. Rev. A 14:36 (1976).
3. L. Hedin, Phys. Rev. 139:A796 (1965).
4. M.S. Hybertsen and S.G. Louie, Phys. Rev. B 43:5390 (1986); R.W. Godby, M. Schlüter, and L.J. Sham, Phys. Rev. Lett. 56:2415.
5. W. Borrmann and P. Fulde, Phys. Rev. B 31:7800 (1985).
6. See e.g. M.R. Norman, D.D. Koelling, and A.J. Freeman, Phys. Rev. B 31:6251 (1985).
7. J. Kübler and A.R. Williams, J. Magn. Magn. Mater. 54-57:603 (1986).
8. P.H. Dederichs, S. Blügel, R. Zeller and H. Akai, to be published.
9. See the article by A.K. McMahan in these proceedings.
10. J. Zaanen, G.A. Sawatzky, and J.W. Allen, Phys. Rev. Lett. 55:148 (1985) and J. Magn. Magn. Mat. 54-57:607 (1986).
11. See the article by B. Brandow in these proceedings.

C

HIGH ENERGY SPECTROSCOPIES
OF NARROW BAND MATERIALS

SECTION C: HIGH ENERGY SPECTROSCOPIES OF NARROW BAND MATERIALS

INTRODUCTION

For the context of this book, the term "high energy spectroscopy" implies spectroscopies involving photons and/or electrons with energies between 10 eV and 10 keV. The advances made in recent years in these spectroscopies, and especially in their interpretation, provided one of the main motivations for this meeting. We felt that the last years had witnessed something of a breakthrough in the use of high energy spectroscopies to study narrow band materials. Even though such spectroscopies have poor resolution and cause a large perturbation, they have in some cases given surprisingly detailed insight into the ground state electronic properties. Here we give a very short historical overview of progress in high energy spectroscopies.

Work on the photoelectron spectroscopy (PS)[1] and x-ray spectroscopy[2] was initiated about the same time, but high resolution was achieved with x rays around 1930[3], whilst for photoelectrons the same stage of development was not achieved until the work of Siegbahn's group between 1956 and 1970[4]. It was in this period that sharp PS core levels, chemical shifts and detailed valence bands were all measured. Since this date, photoemission has replaced x-ray spectroscopy as the major tool for studies of solid state electronic structure. At first almost all of the observations in PS were treated in terms of single particle theories, but around 1967 there was increasing attention to the influence of electron-electron interaction and the many-body response of the system when an electron was removed in XPS, when core holes were created in X-ray absorption, (XAS) or anihilated in x-ray emission (XES). In the next decade the major ideas were generated concerning the relaxation (or changes) in the electron wave functions and energy levels when an electron was removed from a material with interacting electrons. Much work can be related to Anderson's "orthogonality catastrophe"[5] which states that the ground state of a system with an infinite number of fermions is orthogonal to that of the same system in the presence of a scattering potential (in our case a core hole). It was concluded that there should be singularities and asymmetries in some of the thresholds and peaks of XES, XAS and XPS[6-9]. Also shake-up satellites[10] were observed for the first time and the first theoretical formalism for their treatment was given[11,12]. Shake up is a generic term for the satellites which arise in high energy spectra when a "single electron" is excited or relaxed to a lower energy. Shake up arises because the wave functions of the other electrons must change as a result in the change of the field produced by the "active" electron.

It should be mentioned that in the 1970's a whole series of experimental developments took place. Many of these involved the availability of synchrotron radiation, which now completely dominates x-ray absorption spectroscopy. Synchrotron radiation has also had a major impact on valence band photoemission, because it permits use of resonance phenomena to pinpoint the contribution of electrons with a particular site and symmetry to the photoelectron spectrum. However, the synchrotron has not been so all pervading in PS as in XAS. Another important experimental development in this period was the idea of using the monochromators of laboratory XPS equipment to measure Bremsstrahlung Isochromat Spectra[13] (BIS, also known as inverse photoemission) to measure the density of unoccupied states. Another point which should be mentioned in a historical overview is the use of experimental electron spectroscopies, such as Auger spectroscopy[14] and XPS/BIS[15] to measure directly the effective on-site Coulomb correlation energy between valence electrons, U_{eff}. In both cases it has become increasingly obvious with passing years that some theoretical input is necessary to extract U_{eff} if the hybridization of the narrow levels with the other valence bands is not negligible. However these methods remain the most direct experimental measure of U_{eff}. Finally, on the experimental side we should mention the role of commercial ultra-high vacuum equipment and spectrometers which became widely available in the 1970's. These were essential because of the surface sensitivity of many of the spectroscopies and the high reactivity of many narrow band materials.

A series of theoretical papers by Toyozawa and Kotani in 1973-4, treating the spectral shapes when a core hole was created or destroyed in materials with strongly correlated electrons, was a major landmark in studies of narrow band materials[16]. This work represented an incomplete and partially filled valence shell (d or f) as a non-degenerate localized state interacting with the (s) conduction band through s-d or s-f mixing. It was assumed that the first unoccupied d or f level was well above the Fermi level, in the absence of a core hole, but was lowered down below E_F in the presence of a core hole. Kotani and Toyozawa established that this model could represent the basic physics of the spectroscopies and would lead to not only peak and edge asymmetries, but would also lead to satellite peaks in the various spectroscopies in which a core hole was created or destroyed. Kotani and Toyozawa's qualitative comparisons with experiment were the first steps in the development of model treatments which were originally used simply to give a physical explanation of the observed spectral shapes (or spectral functions) in terms of fundamental interactions. However, as a result of the work of Gunnarsson and Schönhammer[17] and others, the process is being reversed and such model treatments are increasingly used to try to deduce the magnitude of the fundamental interactions in narrow band solids from the spectral functions. The original aim here was really to derive, from the models and the spectra, rough parameters as a guide to which approximations would be most sensible in more rigorous theoretical treatments of narrow band materials.

The original aim may have been too modest because it is now recognized that the same parameters derived within the Anderson Hamiltonian from the various high energy spectroscopies can also be used with the same Hamiltonian to explain many rather detailed low energy properties. A particular triumph of these models is the case of the Ce 4f levels, which were thought to be chemically inactive only a couple of decades ago. Spectroscopic methods have contributed most in showing that, on the contrary, the 4f electrons are sufficiently hybridized with the valence bands of many Ce compounds and Ce metal, to provide a significant contribution to the cohesion (or covalency) or Ce solids. As a consequence, the low temperature phase transition[18] and physical properties, such as magnetic moment and dynamic susceptibility[19], can usually be explained applying the same model parameters used to describe the spectroscopic properties related to the 4f electrons. This development is at once both very encouraging and a little disturbing, because there may be a tendency to forget that the whole theory is based on a model and to give more thought to refinement of the parameters in the model than to refinement of the model itself.

Having given a short history of high energy spectroscopies, we should now turn to a few fundamentals. The starting point for all treatment is the Fermi golden rule for transition rates, W_{fi}, between initial and final states, ϕ_i and ϕ_f, as

$$w_{fi} = <\phi_f(N)|t|\phi_i(N)>^2 2\pi/h \lceil(E_f) \tag{1}$$

where t is the transition operator and $\lceil(E_f)$ is the density of states at the final state energy. It is normal to regard the $\phi_{f,i}(N-1)$ as N electron determinants which may be factorized as $\phi_{f,i}(N-1).\phi_j$ where ϕ_j is a one electron wave function[20]. To a first approximation, all the different spectroscopies, except Auger, which are discussed here are one electron spectroscopies. It is thus useful to write equation (1) as

$$w_{fi} = <\phi_f(N-1)|\phi_i(N-1)>^2 <\phi_f|t|\phi_i>^2 2\pi/h \lceil(E_f) \tag{2}$$

Thus if the spectroscopy were photoelectron spectroscopy, ϕ_f would be a continuum wave function and ϕ_i is the wave function of one of the electrons in the system; if the spectroscopy were x-ray absorption, ϕ_i would be a core state and ϕ_f would be an electron in the solid but above the Fermi level. Note that $\phi_f(N-1)$ and $\phi_i(N-1)$ are N-1 electron wave functions of different Hamiltonians, and they are not the same. They differ because of the change of ϕ_i to ϕ_f, which may correspond to complete removal or addition of an electron and may be a very large perturbation. In most cases one makes the "sudden" approximation that this perturbation occurs suddenly. As a result, it is necessary to expand $\phi_i(N-1)$ in terms of all the possible $\phi_f(N-1)$ in order to obtain the spectral shape associated with any one pair of ϕ_i and ϕ_f, i.e. for each pair of ϕ_i and ϕ_f there are many possible final states, $\phi_{fx}(N-1)$ which may be reached with a probability

$$<\phi_{f,x}(N-1)|\phi_i(N-1)>^2 \tag{3}$$

and the consequent observation of many peaks in the spectrum is known as "shake up".

These ideas may also be illustrated with reference to shake up in core level XPS spectra. In 1933 Koopmans had the idea that one particle eigenenergies would be a good approximation to ionization energies[21]. However core level ionization generally leads to a cluster of peaks in the spectrum, spread around the eigenenergy associated with a core level (see figure D1) and this is not in accord with Koopmans' idea. Nevertheless, it is true that, in the sudden approximation the weighted mean of all the cluster of peaks should give the Koopmans' energy[12]. Clearly, each core hole created may lead to one of a variety of final states and the probability for any one is given by equation 3.

The relative importance of the shake-up peaks depends on the size of the perturbation induced by the spectroscopy and the relaxation of the "spectator" electrons. The perturbation is largest for spectroscopies where an electron is added or removed, such as BIS or photoemission, and is smallest when the number of electrons and core holes is conserved, as in x-ray emission, as a result of migration of a core hole from a deep to a less-deep core level.

In descriptions of the low energy properties it is usual to speak of quasiparticles, which consist of electrons plus their disturbance of the system. It must be stressed that a technique like photoelectron spectroscopy measures the one-electron (not quasi-particle) Green's function and the spectral distribution hereof yields information on the nature of the screening cloud accompanying the electrons in the quasiparticle.

For instance, in a very simple system one <u>might</u> be able to describe a quasiparticle consisting of an electron plus its exchange correlation hole. In this case, when the electron is removed the spectral distribution must somehow reflect the exchange correlation hole that accompanied the electron in the quasiparticle. Another example is where the screening of an electron takes place via a lattice distortion, in which case the quasiparticle is a small polaron and the photoelectron spectrum would be accompanied by phonon sidebands whose amplitudes reflect the degree and shape of the distortion. For metallic systems there would still be a cut-off in the photoelectron spectrum corresponding to the Fermi energy. However, the spectral weight could be vanishingly small, making it look like a semiconductor.

We continue with a phenomenological discussion of observations for core level XPS (figure 1b-d). Figure 1b shows the 2s level spectra for Al. The parent XPS peaks are accompanied by strong satellites due to creation of plasmons (hw_p). Plasmons are collective excitations of an electron gas[22] whose energy in Al is ≈ 15 eV. Because their energy changes with momentum, $\underline{k}$, the energy taken from the photoelectron on plasmon creation varies, and the plasmon satellites have a skewed shape. Surface plasmon creation, costing ≈ 10 eV is also observed. Not all of the satellite intensity observed here

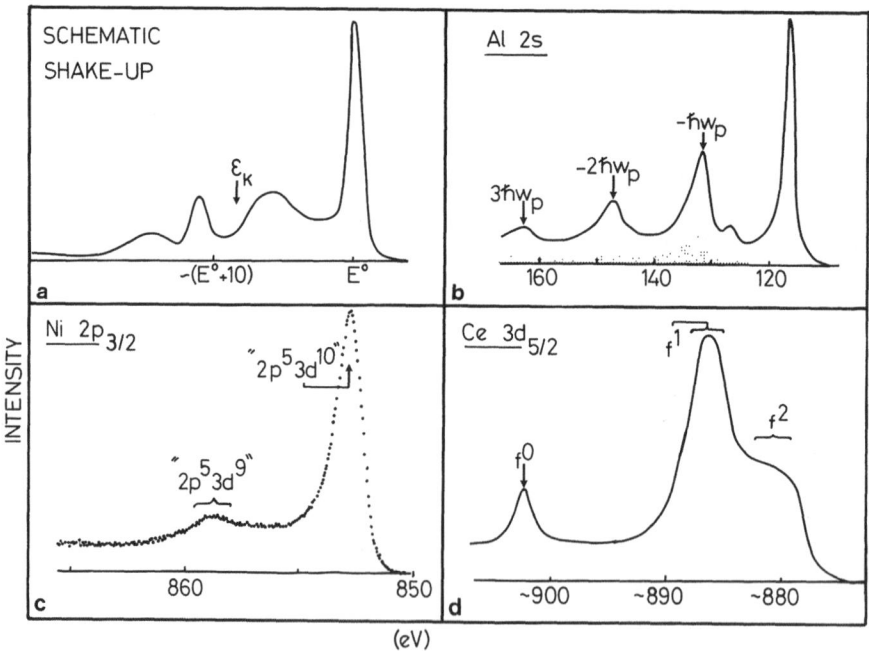

Fig. 1.   a) Schematic diagram of shake up around a core level energy in XPS.
b) Al 2s XPS spectra, with associated satellites.
c) Ni $2p_{3/2}$ XPS spectrum with the usual peak designations.
d) Schematic diagram of Ce $3d_{5/2}$ XPS peaks.

is shake-up because the photoelectrons may also excite plasmons via inelastic scattering. The intrinsic part of the losses is given schematically by the shaded area in figure 1b. Note that the Koopmans energy, or centres of gravity of the intrinsic contributions are 5-10 eV to the high binding energy sides of the parent peaks.

The other examples of shake-up, given in figures 1c and 1d, are the spectra of Ni $2p_{3/2}$ and Ce $3d_{5/2}$ core levels. The Ce spectrum is partially schematic to suppress the complications arising because the 3d core levels consist of a doublet. The shake up peaks in figures 1c and 1d are also manifestations of equation 3 which stated that when a given core hole is created in an N electron system there is a statistical probability of reaching many different, N-1 electron, final states. To interpret these spectra, and in particular to draw conclusions about the ground state, one must resort to a simplified model. In the cases of Ni and Ce there is much evidence that the peaks correspond to transitions to states in which a single, local configuration dominates. For instance, for Ni, the configurations are $2p^5 3d^{10}$ and $2p^5 3d^9$ and the charge is conserved by changes in the conduction band occupation. The models normally used for these systems are built on the simplifying assumption that whilst there are many possible configurations (or basis states) that may mix to form the initial and final states, only a few local configurations (d^9 and d^{10} for Ni and f^0, f^1 and f^2 for Ce) are important and that these may be considered separately. As Martensson points out in his chapter, there are also strong satellites in systems without narrow bands and it is not clear that they may always be neglected.

The final stage in this section is given over to a phenomenological description of the development of the XPS/BIS valence band spectra with increasing localization of the valence states. This is illustrated in figure 2. Starting with a free-electron gas, the density of electronic states in such a material should increase with $E^{1/2}$. Al is often considered as a nearly free-electron gas and its computed density of states (DOS, strictly speaking "density of eigenstates") has some definite peaks and valleys superimposed on the free-electron-like curve. The actual XPS/BIS spectra strongly resemble the computed DOS. The main differences are due to the plasmon-loss contribution (subtracted from figure 2a) and the one particle transition matrix elements which enhance the contribution of some states to the spectra[29]. The small differences between the computed DOS peak energies and those observed may be attributed partly to the XPS/BIS matrix elements, but also to the energy dependence of the exchange-correlation interaction. This latter, also known as the self energy in the density functional formalism, is a fundamental effect. The binding energy of electrons in a solid may be regarded as being stabilized by of the order of 10 eV as a result of interaction with their own exchange-correlation hole. This stabilization is energy dependent in that if the kinetic energy of an electron in a solid is large, the electron becomes decoupled from its exchange correlation hole and the stabilization disappears. Experimentally one refers all electron energies to the Fermi level where the stabilization V_{XC}, has a particular value. However as V_{XC} is not always the same as at E_F, this results in shifts between the experimental and computed peak energies[30]. This can only be corrected by a full calculation of V_{XC} (or the energy dependence of the self energy in computation schemes based on the density functional.)

In the transition metals, the metal d wave functions peak inside the Wigner Seitz radii and the overlap between the wave functions centred on different sites is weaker than for Al. The d-states thus form a well defined

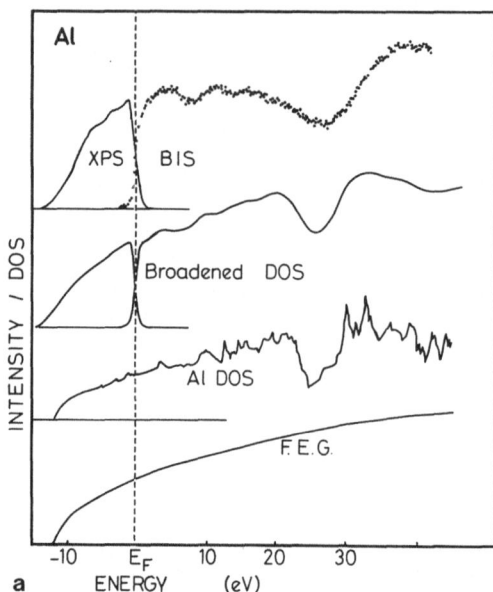

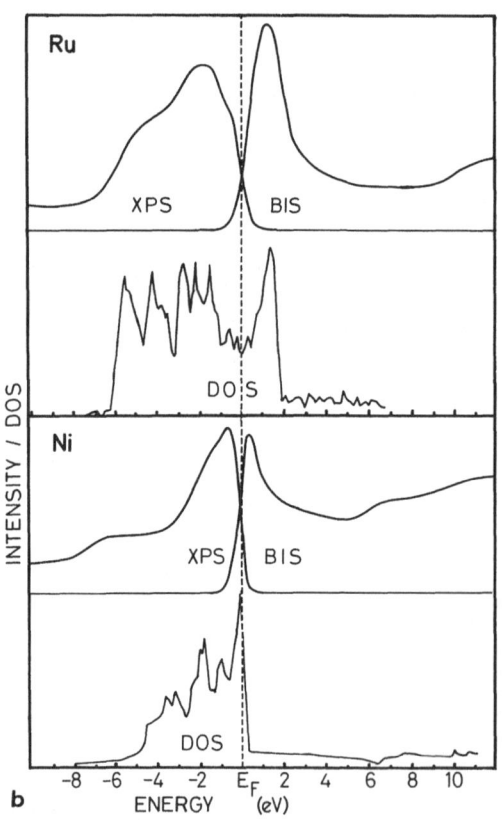

Fig. 2. Illustration of the effect on XPS/BIS spectra, of increased localization of valence states.
a) Al, using data from reference 23. Bottom curve: free-electron-gas. Next: compound DOS. Next: DOS fold with a lifetime broadening. Top Curves: XPS/BIS.
b) Ru and Ni. Lower curves: DOS. Top curves: XPS/BIS. Data from reference 24 (Ru) and 25 (Ni).

band of states, usually between 4 and 10 volts wide, with sp states above and below the d band[25,31]. The XPS/BIS spectra generally reflect the one particle DOS closely, as illustrated for Ru in figure 2b.

The band width, W, of the transition metals decreases and the effective on-site Coulomb correlation energy, U_{eff}, generally increases from left to right across the periodic table. Ni has the smallest bandwidth and the largest U_{eff} of all the transition metals[32]. This, combined with its nearly filled d band, results in a strong distortion of the observed XPS spectrum with respect to the independent-particle DOS. A satellite, corresponding quite closely to production of sites with a local Ni d^8 configuration, is pushed out of the main band to about 6 eV below E_F by the strong d-d correlation energy. In addition, the Ni 3d band is narrower in the photoelectron spectrum than in the single particle DOS.[26]

The BIS spectra of Ni and all the transition metals correspond closely to the calculated DOS,[25] especially if one takes into account the BIS matrix elements, which generally enhance the contribution of the sp states for the 3d transition metals. Until now no satellites comparable to those caused by correlation in the XPS valence band of Ni have been observed and a search for such effects is more profitable in transition metal compounds, where the ratio of U/W is larger. Strong satellites are well known, for instance for CuO, NiO, and the halides of Cu and Ni, for instance.

The XPS/BIS spectrum of Tb (Fig. 3) is fairly typical of those for the rare earths. The main peaks in the XPS correspond, for Tb, to excitations from the f^8 ground state to different f^7 final states. Similarly, the BIS peaks on the right hand side correspond to excitations from the f^8 ground state to different f^9 final states. Excitations to excited f^8 states are not seen in XPS/BIS but it is clear from the complicated structure in the spectra

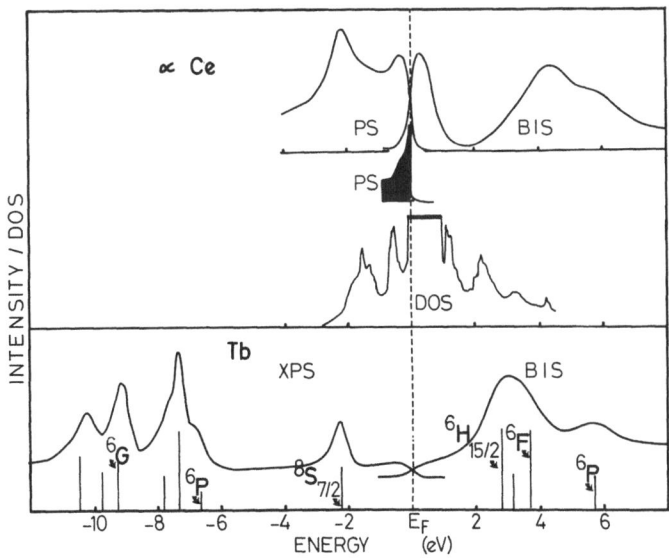

Fig. 3.   Ce and Tb valence bands.
Lower curve: XPS/BIS of Tb (Adapted from ref. 15). Next: αCe DOS from reference 27. Top: αCe XPS/BIS adapted from references 15,18. The shaded curve is a high resolution UPS spectrum of the region near E_F.

that multiplet splittings in the rare earths are very large. We may also obtain a good estimate of the effective Coulomb interaction in Tb from figure 3. U_{eff} is defined as

$$U_{eff} = 1P - ea$$

where 1P is the ionization potential and ea is the electron affinity. The lowest ionization energy leads to the $S_{7/2}$ f^7 state and the lowest energy electron addition leads to the $^6H_{15/2}$ f^9 state. The difference between these two peaks in the XPS/BIS is ≈ 5 eV, and that is U_{eff} for Tb.

As discussed in the introduction to the book, and the section on low energy properties of narrow band materials, a complicated situation arises when the energy of a change in number of f electrons becomes comparable to the integrals mixing the localized f-levels with the delocalized conduction bands. The simplest picture is one of mixed valence, where the f levels straddle E_F, and the f-count fluctuates. This picture is not always adequate because it is now well known that for Ce, if the $4f^1$ eigenenergy is ≈ 2 eV, a hybridization integral of ≈ 20 meV may produce a strong singularity in the density of quasiparticles near E_F. Further, a hybridization integral of only one or two hundred meV may spread throughout the whole valence band, the spectrum associated with a single f state. There is much evidence for this point from resonance photoemission studies of the Ce valence band[17-19,33]. We illustrate the point with the spectra and density of states of αCe.

A one-particle density of states of αCe shows a band of 4f states around E_F which is about 1 eV wide[27]. This band can accommodate 14 electrons and does not take the electron-electron interactions fully into account. Thus, the PS/BIS spectra show peaks at ≈ -2.3 eV in PS and $\approx +4.5$ eV in BIS for which the descriptions $f^1 \rightarrow f^0$ and $f^1 \rightarrow f^2$ are a reasonable approximation. Peaks are also found at E_F and high resolution photoelectron experiments (shaded peak in figure 3) show this peak to be very sharp. This peak is related to the high density of quasi-particle states at E_F indicated by low temperature experiments on αCe, but as described above its <u>weight</u> in the photoelectron may be modified not only by single particle matrix elements, but also by the spread of spectra weight to other energies because again, stated above, the photoelectron spectrum reflects not only the quasi-particle energies, and the photoelectron spectrum also contains information on the screening cloud accompanying the electrons in the quasi particle.

REFERENCES

1. For a review of early photoelectron spectroscopy, see J.G. Jenkin, R.C.G. Leckey and J. Liesegang, <u>J</u>. <u>Electron</u> <u>Spectroscopy</u> 12:1 (1977).
2. For a review of early X-ray spectroscopy see A.H. Compton and S.K. Allison, in: "X-rays in Theory and Experiment", Publ. van Nostrand, Princetown, (1935).
3. H.H. Johann, <u>Z</u>. <u>f</u>. <u>Phys</u>. 69:185 (1931);
    J. Johansson, <u>Z</u>. <u>f</u>. <u>Phys</u>. 82:507 (1933).
4. See e.g. K. Siegbahn, C. Nordling, A. Fahlman, R. Nordberg, K. Hamrin, J. Hedman, G. Johansson, T. Begmark, S.-E. Karlsson, I. Lindgren, and B. Lindberg, in: "ESCA-Atomic, Molecular and Solid State Structure Studies by Means of Electron Spectroscopy", Nova Acta Regiae Soc. Sci. Uppsaliensis 20, (1967);
    K. Siegbahn, C. Nordling, G. Johansson, J. Hedman, P.F. Hedén, K. Hamrin, U. Gelius, T. Bergmark, L.O. Werme, R. Manne and Y. Baer, in: "ESCA Applied to Free Molecules", N. Holland, Amsterdam, (1971);
    K. Siegbahn, <u>J</u>. <u>Electron</u> <u>Spectroscopy</u> 5:1 (1974).
5. P.W. Anderson, <u>Phys</u>. <u>Rev</u>. <u>Lett</u>. 18:1049 (1967);
    <u>Phys</u>. <u>Rev</u>. 164:352 (1967).

6.   G.D. Mahan, Phys. Rev. 163:612 (1967).

7.   P. Nozieres and C.T. de Dominicis, Phys. Rev. 178:1097 (1969).

8.   G.A. Ausman and A.J. Glick, Phys. Rev. 183:687 (1969).

9.   S. Doniach and M. Sunjic, J. Phys. C3:285 (1970).

10. T.A. Carlson and M.O. Krause, Phys. Rev. 140:A1057 (1975);
     T.A. Carlson, C.W. Nestor, Jr., T.C. Tucker and F.B. Malik, Phys.
     Rev. 169:27 91968).

11. B.I. Lundqvist, Phys. Kondens. Mater. 6:193 (1967).

12. R. Manne and T. Åberg, Chem. Phys. Lett. 7:283 (1970).

13. J.K. Lang, and Y. Baer, Rev. Sci. Instr. 50:221 (1979).

14. M. Cini, Solid State Commun. 24:681 (1977) and Phys. Rev. B15:2788
     (1978); G.A. Sawatzky, Phys. Rev. Lett. 34:504 (1977).

15. P.A. Cox, J.K. Lang, and Y. Baer, J. Phys. F11:113, 121 (1981).

16. A. Kotani and Y. Toyozawa, J. Phys. Soc. Japan 35:1073,1082 (1973);
     37:912 (1974).

17. K. Schönhammer and O. Gunnarsson, Solid State Commun. 23:691 (1977);
     26:147,399 (1978); J.C. Fuggle, M. Campagna, Z. Zolnierek, R. Lasser
     and A. Platau, Phys. Rev. Lett. 45:1597 (1980);
     O. Gunnarsson and K. Schönhammer, Phys. Rev. Lett. 50:604 (1983);
     Phys. Rev. B27:4315 (1983).

18. J.W. Allen and R.M. Martin, Phys. Rev. Lett. 49:1106 (1982);
     J.W. Allen, S.J. Oh, O. Gunnarsson, K. Schönhammer, M.P. Maple, M.S.
     Torikachvili and I. Lindau, Adv. in Phys. 35:275 (1986) and references
     therein.

19. O. Gunnarsson, K. Schönhammer, J.C. Fuggle, F.U. Hillebrecht, J.-M.
     Esteva, R.C. Karnatak and B. Hillebrand, Phys. Rev. B28:7330 (1983).

20. One speaks of complete breakdown of the one electron wave function
     when this factorization is not sensible. These cases are extremely
     interesting, but less relevant here.

21. J. Koopmans, Physica 1:104 (1933).

22. see e.g. p ff in C. Kittel "Introduction to Solid State Physics",
     (5th Edition, Publ. Wiley, New York, 1976).

23. H.J.W.M. Hoekstra, W. Speier, R. Zeller and J.C. Fuggle, Phys. Rev.
     B34:5177 (1986). Y. Baer and G. Busch, Phys. Rev. Lett. 30:280 (1973).

24. J.C. Fuggle, p. 273 in: "Laboratory Methods in Photoelectron Spectro-
     scopy", Ed. D. Briggs, Publ. Heyden, London (1978).

25. W. Speier, J.C. Fuggle, R. Zeller, B. Ackermann, K. Szat, F.U.
     Hillebrecht and M. Campagna, Phys. Rev. B 30:6921 (1984).

26. J.C. Fuggle, F.U. Hillebrecht, R. Zeller, Z. Zolnierek, P.A. Bennett
     and Ch. Freiburg, Phys. Rev. B27:2145 (1983) and references therein.

27. W.E. Pickett, A.J. Freeman and D.D. Koelling, Phys. Rev. B23:1266
     (1981); see also P. Podloucky and D. Glötzel ibid. 27:3390 (1983).

28. N. Mårtensson, B. Reihl and R.D. Parks, Solid State Commun. 41:573
     (1983); E. Wuilloud, H.R. Moser, W.-D. Schneider and Y. Baer, Phys.
     Rev. B28:7354 (1983); F. Patthey, B. Delley, W.-D. Schnieder and Y.
     Baer, Phys. Rev. Lett. 55:1518 (1985).

29. W. Speier, J.C. Fuggle, P. Durham, R. Zeller, R.J. Blake and P. Sterne,
     J. Phys. C. (1988) in press.

30. W. Speier, R. Zeller and J.C. Fuggle, Phys. Rev. B32:3597 (1985).

31. V.L. Moruzzi, J.F. Janak and A.R. Williams,"Calculated Electronic
     Properties of Metals, Pergamon Press, New York (1978).

32. Actually U_{eff} may be larger in Mn because of the large multiplet split-
     tings in Mn d^5 states.

33. These resonance phenomena are related to the Fano effect which produces
     strong variations in partial photoelectron cross sections near a core
     excitation threshold; U. Fano Phys. Rev. 124:1866 (1961). This effect
     is increasingly used in conjuction with photoemission induced by syn-
     chrotron radiation, to accentuate the spectral weight of a single
     site and symmetry selected contribution to the total density of (quasi-
     particle) states.

HIGH ENERGY SPECTROSCOPIES

J. W. Allen*

Xerox Palo Alto Research Center
Palo Alto, CA 94304, USA

In the past ten years the so-called high energy spectroscopies have had a major impact on the understanding of the electronic structure of narrow-band systems. These spectroscopies have been used to determine the occupations, valence states and spectral weight distributions of the narrow band electronic states in question, 3d for transition metals, 4f for rare earths, or 5f for actinides. It is found that the spectra show systematic differences from the results of solid state local density functional calculations and thereby reveal the presence of the large Coulomb interactions in these systems. For pure transition metals[1,2] and some metallic transition metal compounds,[2,3] the spectra have been modeled with some success by the Hubbard Hamiltonian. For the rare earths[4] and certain insulating transition metal compounds,[2,5-11] the spectra can be described quite well by the impurity Anderson Hamiltonian, and if the Hamiltonian parameters thus obtained are used to evaluate quantities of significance for the low energy properties of the system, such as characteristic energies for loss of magnetic moments,[4] or for magnetic ordering[9], the results are in reasonable agreement with the values obtained directly from low energy experiments. Thus one has a unified picture of the low energy properties that permit an impurity model description and of the underlying electronic structure giving rise to these properties. In most cases the picture that has emerged from this program is quite different[4,9] from that which was accepted before the addition of the spectroscopic information. It will be interesting to see whether this occurs also for the high T_c copper oxide superconductors.

The high-energy spectroscopies are generally taken to include photoemission, inverse photoemission, Auger, and x-ray absorption and emission. The discussion here will focus mostly on the first two of these, which are also termed electron spectroscopies. In photoemission spectroscopy (PES) monochromatic photons impinge on the sample and the kinetic energy spectrum of the resulting photoelectrons is measured. Inverse photoemission is the time reversed process, in which monochromatic electrons incident on the sample undergo radiative transitions to unoccupied states and the resulting Bremsstrahlung photon spectrum is measured. This experiment is often called Bremsstrahlung isochromat spectroscopy (BIS) because it is common to fix the photon energy detected and to vary the incident electron kinetic energy.

For some workers there persists a negative connotation to the term high-energy, reflecting a feeling that the probe disturbs the system so as to preclude gaining useful information about the delicate phenomena that make narrow band systems so interesting. This is clearly a legitimate concern when one tries to obtain information about the valence electrons from spectroscopies in which core holes are created in the final states, either photoemission or photoabsorption. For the core spectrum to carry such information there must be an interaction between the core hole and the valence electrons. Generally this interaction is modeled in the simplest way that yields a description of the spectrum, leaving the worry that other effects of the core hole, such as parameter renormalization, have gone undetected.

For the case of valence band PES and conduction band BIS of the narrow band electrons themselves, the situation is quite different. If the kinetic energy of the out-going or in-coming electron is sufficiently large that the electron does not interact with the system after or before the photon event, respectively, then one measures the ionization and affinity energies of the narrow band electrons, i.e., the energies to remove and add these electrons from the system. It is important to be aware that the theorist has a general tool for describing the ionization and affinity spectrum of a many-body system, this being the single-particle Green's function, where the term 'single-particle' is not to be confused with 'non-interacting'. It is to be emphasized that for an interacting many-electron system, the ionization and affinity energies are precisely the single-particle, or one-electron spectral quantities that one is allowed to know, within the usual interpretation of quantum mechanics. It is not possible to follow the behavior of one electron without either removing it or adding it, and there is no meaning to the question one occasionally hears, or reads, 'what is the energy of, e.g., the f-electron, in the ground state?' Thus the real issue of importance for electron spectroscopy of the narrow band electrons is not the largeness of the photon energy, but the adequacy of the resolution for the study being done.

The large impact claimed above for the experimental spectroscopy work can be traced in part to steady advances in the technique. One of these advances is the use of the synchrotron as a photon source. With the availability of tunable high-intensity radiation, photoemission experiments have made use of the photon-energy dependence of cross-sections in general, and of resonances in the cross-sections in particular, to separate overlapping contributions to valence bands. For example, these techniques have been used to determine the 4f spectral weight in cerium materials, and to identify split-off 3d states in transition metal spectra. There has been a steady push[12] toward higher resolution in this work, with values between 140 and 400 meV being common. It seems likely that the new high-intensity beam lines using insertion devices will enable resolutions an order of magnitude better. It also seems likely[13] that studies with rare earth, actinide or transition metal impurities as dilute as 0.1% will be possible. In x-ray absorption spectroscopy the synchrotron has made it relatively easy to make measurements with good signal to noise ratio in reasonable periods of time. Thus data have been taken on large numbers of materials chosen to reveal systematic behaviors, and on particular materials with variable temperature and pressure. A second advance has been the increased resolution, of order 10 to 20 meV, of laboratory ultraviolet photoemission, using the helium lamp as a source. For situations where the photon energies available, 20eV and 40eV, suffice, very interesting and beautiful results[14,15] have been obtained, whetting the appetite for the high resolution work with tunable photon energy that is anticipated from the synchrotron. Third, inverse photoemission has been extremely important because the effects of Coulomb interactions cannot be adequately determined without the combined electron addition and removal spectra.

Spectroscopy cannot succeed without theory for interpreting the data. For many-electron systems two important advances can be noted. One[4] is the development of techniques for calculating the ground state properties and the photoemission and absorption spectra of the Anderson impurity Hamiltonian. This Hamiltonian describes a local orbital of degeneracy N, binding energy ε, and Coulomb interaction U, hybridized to a continuum via a matrix element V. An exciting aspect of this model Hamiltonian is its relaxation and screening properties, which are discussed further below, and which seem to be observable. The understanding of the modifications of these results for a lattice version of the Hamiltonian remains scant. Of equal importance is the density-functional theory, which is capable, in principle, of yielding certain average ground state properties exactly. The eigenvalues are not a priori physically observable quantities, but in systems with weak Coulomb interactions they often are in good agreement with measured electron spectra, and even in systems with strong Coulomb interactions, predict Fermi surfaces much like those observed. The latter result presumably reflects the tendency for the Fermi surface geometry to be determined by the Fermi surface phase shifts, which are rendered somewhat immune to Coulomb interactions by Fermi surface sum rules. Although the density-functional eigenvalues do not yield a good description of the ionization and affinity spectrum of strongly interacting systems, the theory has the great advantage of not being a model theory. It is then very exciting to see a few recent efforts[16-19] to simulate excited states and thereby to obtain from first principles the parameters of the Anderson Hamiltonian. These parameters can then be compared with the ones obtained by fitting an experimental spectrum.

The aspect of the Anderson model which is perhaps most intriguing is the tendency of the local orbital to relax, after the removal or addition of an electron, to an occupation as near to that of the ground state as is possible. This is accomplished by the hybridization term in the Hamiltonian, which leads to exchange of holes and electrons with the continuum. This relaxation is manifested in the local-orbital single-particle spectrum by a peak which is a derived property of the model. For a metal, e.g. a cerium intermetallic, this peak appears at the Fermi energy, and corresponds to final states with the local orbital

occupation exactly the same as in the ground state. It is called the Kondo resonance, and its width and weight are related to the Kondo temperature T_K, which is the energy scale on which the local orbital electron's magnetic moment is quenched. For an insulator, e.g., NiO, an analagous peak appears at the top of the valence band, and corresponds to final states in which the local orbital occupation is close to, but somewhat less than, that of the ground state. In this case, the peak has no special name and there appears to be no special energy scale. It is important to realize that the Kondo resonance is a particular example of a relaxation behavior which is a general feature of the Anderson Hamiltonian. Of course, it is a very special example because it lies at the Fermi level (apart from extra structure such as spin-orbit sidebands[4]) and hence is related to ground state properties, and to Fermi surface sum rules. The relaxed peak also has a conceptual relation to the results of a density functional calculation, in that the eigenvalues obtained therein are found from a potential derived with the ground state occupations of orbitals.

For the case of an insulator a simple example of the relaxed peak in photoemission can be constructed with a cluster model. Such an example may be of special interest at the moment because of possible application of this type of picture to the superconducting copper oxides. Consider an $(MO_6)^{10-}$ cluster, as occurs in NiO, with octahedral coordination of oxygen atoms O around a formally divalent metal atom M, and introduce basis states $|1> = d^n p^6$ and $|2> = d^{n+1} p^5$. Here 'd' and 'p' are the 3d and 2p orbitals of the M and O atoms, respectively, and all crystal field and multiplet splittings are ignored. For NiO, n=8, and for CuO, n=9. The two configurations are coupled by a hybridization matrix element V and the second is taken to lie an energy E above the first, which is the purely ionic state. The Hamiltonian is then represented as a 2x2 matrix with elements $h_{11}=0$, $h_{22}=E$, and $h_{12}=h_{21}=V$. Call this the Hamiltonian of the N electron system. Similarly, states with one electron removed, i.e., for the N-1 electron system, are $|2> = d^{n-1} p^6$ and $|1> = d^n p^5$. These are coupled by a hybridization matrix element V' and the $d^{n-1} p^6$ configuration is taken to lie highest by an energy E'. It need not be that the $d^n p^5$ state lies lowest, but details of resonant photoemission behavior provide evidence that this is the case for the Ni and Cu transition metal compounds studied in detail thus far. The Hamiltonian matrix for these states is then $h_{11}=0$, $h_{22}=E'$ and $h_{12}=h_{21}=V'$. The ground state of the N electron system is $|G,N> = a|d^n p^6> + b|d^{n-1} p^5>$ and the two possible states of the N-1 electron system are $|i,N-1> = a_i |d^n p^5> + b_i |d^{n-1} p^6>$, with i=1,2. The d-photoemission intensity M_i into each of the states $|i,N-1>$ is proportional to the square of their overlap with the state produced by removing one d-electron from $|G,N>$, so that $M_i = (ab_i + ba_i)^2 = a^2 b_i^2 + 2ab_i ba_i + b^2 a_i^2$. Note the appearance of the interference term in M_i.

The weight in the lowest energy line depends on V. If V=V'=0, there is no mixing of the wavefunctions, so $a=a_1=1$, $b=b_1=0$, $a_2=0$ $b_2=1$, $M_1=0$ and $M_2=1$. That is, all the photoemission intensity goes into the higher energy line, and none into the relaxed peak. Assume now for simplicity that E=E'=8 and V=V'=3, so that the Hamiltonian is the same for the N and N-1 electron systems. The eigenvectors of the Hamiltonian are easily found to be $(1/\sqrt{10})(-3,1)$ and $1/\sqrt{10}(1,3)$ with eigenvalues of -1 and 9 respectively, leading to the identifications $a=a_1=-3/\sqrt{10}$, $b=b_1=1/\sqrt{10}$, $a_2=1/\sqrt{10}$ and $b_2=3/\sqrt{10}$. Substituting into the expression for M_i gives $M_1=(1/100)(9+9+18)=36/100$ and $M_2=(1/100)(81+1-18)=64/100$. For these choices of E and V, the mixings in the wavefunctions are 10% and 90%, but the distribution of intensity is quite different, 36% into the lowest energy line, which is the relaxed peak, and 64% into the higher energy line. If E=E'=8 and V=V'=$2\sqrt{5}$, then it is easily verified that the Hamiltonian eigenvalues are -2 and 10, with $a=a_1=-\sqrt{5}/\sqrt{6}$, $b=b_1=1/\sqrt{6}$, $a_2=1/\sqrt{6}$ and $b_2=\sqrt{5}/\sqrt{6}$. In this case $M_1=20/36$ and $M_2=16/36$, so that a modest 17%/83% mix in the wavefunctions transfers more than half, 56%, of the photoemission intensity into the relaxed peak.

The rapid transfer of weight into the lowest energy line with increasing V is due to the interference term in the expression for M_i, which reflects the fact that the phasings in the wavefunctions for the ground states of the N and N-1 electron systems are the same. It is a rather general spectroscopic principle that an amount of weight disproportionate to the wavefunction mixing is transfered to the lowest energy line. This principle also operates in core level photoemission and photoabsorption in cases where hybridization mixes configurations in both the initial and final states, so that the intensities of lines corresponding to different valence states does not provide an accurate measure of the wavefunction mixings when V is large.

It should be noted that the picture of this example has met disagreement[20-23] from workers convinced that in CeF_4 and CeO_2, Ce must be tetravalent, or, presumably, that in NiO, Ni must be divalent. Often this opposition results from interpreting experimental spectra without taking into account the detailed properties of the model, such as its energetics, or the effect of quantum interference on line intensities. Another motivation for opposition to the model seems to be the intuitive conviction that covalency can be sharply distinguished from mixed valence, even though both are understood to arise from the same mechanism, hybridization between band and local electrons states. In this view, the filled oxygen band of CeO_2 would be regarded as 'having f-character' due to covalency, but it would be asserted that there is a 'local f-orbital' which is empty, implying tetravalence. In an Anderson model description, as was actually used in one paper whose text expressed the covalency view, then it is certainly the local orbital spectral weight that is mixed into the oxygen band. This weight corresponds to f^1p^5 states in the ground state and to a nonzero occupation of the local orbital, about 0.5 electrons.

As described in more detail elsewhere,[24] one can formulate an alternative description, which is of the covalent sort, by introducing *hybridized* p/f orbitals, say ph and fh, in terms of which the N electron ground state could be written as ph^6fh^0. The analog for NiO would be ph^6dh^8. For an experiment in which these p/f or p/d bonds are not broken into their atomic components, the ph/fh or ph/dh basis will provide a useful description. In NiO the excitation of spin waves or possibly 3d optical excitons would be of this type. But a probe sensitive to the atomic local orbital, such as photoemission, will break the bonds and will reveal the partial occupation due to covalency, to

the extent permitted by the kinds of quantum interference effects described above. Whether this partial occupancy is called mixed valence or covalency is not of central importance, but it might be reasonable to agree to reserve the former term for systems with low energy charge fluctuations.[24]

Looking to the future there are many problems and opportunities available in the general areas outlined above. At this particular instant, one of the most exciting is to see if the Anderson Hamiltonian picture that appears to work for Cu halides and NiO can be extended[25,26] to data[27,28] for CuO, Cu_2O and the high T_c superconducting copper oxides, and whether[29] the correlated electron aspects of this picture have any relation to the mechanism of the superconductivity. Along the same lines, the applicability of the Anderson Hamiltonian to metallic transition metal compounds like NiAs[30], or even NiS[31], appears to be difficult, and remains to be established also for the metallic actinide materials, especially the heavy Fermion ones. The spectral differences between the impurity and the lattice version of the Anderson Hamiltonian have yet to be established either theoretically or experimentally. For cerium materials, and by extension, others, there is continuing debate as to the necessity of agumenting the Hamiltonian with a Coulomb interaction between electrons in the local orbital and ones in the continuum states. It is also found that the local orbital parameters needed to fit a core level spectrum are similar to but not identical with those which produce the best fits of valence and conduction band spectra. It is not clear yet whether this problem is due to the effect of the core hole in renormalizing parameters, or whether it signals some more important deficiency of the model. Finally, the precise relation of the density functional theory to model Hamiltonian theories, or to experimental spectra remains to be clarified, although efforts in this direction were noted above.

* Permanent address after Sept. 1, 1987: Physics Dept., Univ. of Michigan, Ann Arbor, MI. 48109-1120.

References

1. D. R. Penn, Phys. Rev. Letters, 42:921 (1979).

2. L. C. Davis, J. Appl. Phys., 59:R25 (1986).

3. O. Bisi, C. Calandra, U. Del Penninko, P. Sassaroli and S. Valeri, Phys. Rev. B, 30:5696 (1984).

4. J. W. Allen, S.-J. Oh, O. Gunnarsson, K. Schonhammer, M. B. Maple, M. S. Torikachvili and I. Lindau, Adv. in Physics, 35:275 (1986).

5. G. van der Laan, C. Westra, C. Haas and G.A. Sawatzky, Phys. Rev. B, 23:4369 (1981).

6. G. van der Laan, Solid State Commun., 42:165 (1982).

7. G. A. Sawatzky, in "Studies in Inorganic Chemistry," Volume 3 (Elsevier, Amsterdam, 1983) page 3.

8. A. Fujimori and F. Minami, Phys. Rev. B, 30:957 (1984).

9. G. A. Sawatzky and J. W. Allen, Phys. Rev. Letters, 53:2339 (1984).

10. J. Zaanen, G. A. Sawatzky and J. W. Allen, Phys. Rev. Letters, 55:418 (1985)

11. J. Zaanen, G. A. Sawatzky and J. W. Allen, J. Magn. and Mag. Mat., 54-57:607 (1986).

12. D. M. Wieliczka, C. G. Olson and D. W. Lynch, Phys. Rev. B, 29:3028 (1984).

13. J.-S. Kang, J. W. Allen, M. B. Maple, M.S. Torikachvili, B. Pate, W. Ellis, and I. Lindau, submitted for publication.

14. F. Patthey, B. Delley, W.-D. Schneider and Y. Baer, Phys. Rev. Lett., 55:1518 (1985).

15. F. Patthey. W.-D. Schneider, Y. Baer and B. Delley, Phys. Rev. Lett., 55:1518 (1985).

16. M. R. Norman, D. D. Koelling, A. J. Freeman, H. J. F. Jansen, B. I. Min, T. Oguchi, Ling Ye, Phys. Rev. Lett., 53:1673 (1984).

17. M. R. Norman and D. D. Koelling, Phys. Rev. B, 31:6251 (1985).

18. M. R. Norman, Phys. Rev. B, 31:6261 (1985).

19. A. McMahan and R. M. Martin, unpublished work on CeO2.

20. P. Wachter in "Valence Instabilities," edited by P. Wachter and H. Boppart (North-Holland, Amsterdam, 1982) p. 145.

21. E. Wuilloud, B. Delley, W.-D. Schneider and Y. Baer, Phys. Rev. Lett., 53:202 (1984).

22. G. Kalkowski, C. Laubschat, W. D. Brewer, E.V. Sampathkumaran, M. Domke and G. Kaindl, Phys. Rev. B, 32:2717 (1985).

23. G. Kaindl, G. K. Wertheim, G. Schmiester and E. V. Sampathkumaran, Phys. Rev. Letters, 58:606 (1987).

24. J. W. Allen, J. Magn. and Mag. Mat., 47-48:168 (1985).

25. A. Fujimori, E. Takayama-Muromachi, Y. Uchida and B. Okai, Phys. Rev., 1987, to be published.

26. Zhi-xun Shen, J.W. Allen, J.J. Yeh, J.-S. Kang, W. Ellis, W. Spicer, I. Lindau, M.B. Maple, Y.D. Dalichaouch, M.S. Torikachvili and J.Z. Sun, Phys. Rev. B, to be published.

27. M. R. Thuler, R. L. Benbow and Z. Hurych, <u>Phys. Rev. B</u>, 26:669 (1982).

28. S.-J. Oh, J.W. Allen, I. Lindau, and J.C. Mikkelsen, Jr., <u>Phys. Rev. B</u>,26: 4845 (1982).

29. P. W. Anderson, <u>Science</u>, 235: 1196 (1987); P.W. Anderson, G. Baskaran, Z. Zou and T. Hsu, <u>Phys. Rev. Letters</u>, 58: 2790 (1987).

30. W. P. Ellis, R.C. Albers, J.W. Allen, Y. Lassailly, J.-S. Kang, B.B. Pate and I. Lindau, <u>Solid State Commun.</u>, 62: 591 (1987).

31. J. Zaanen, G.A. Sawatzky and J.W. Allen, <u>J. Magn. and Mag. Mat.</u>, 54-57: 607 (1986). Unpublished analysis shows that it is difficult to fit the core level spectra with the same parameters as for the PES-BIS spectra reported in this paper. Also the kink in the least binding energy valence band peak cannot be reproduced with the impurity theory.

HIGH ENERGY SPECTROSCOPIES AS A PROBE OF HYBRIDIZATION, COULOMB CORRELATION

ENERGIES AND OCCUPANCIES IN NARROW BAND SYSTEMS: DISCUSSION PAPER

J.C. Fuggle

Research Institute for Materials, University of Nijmegen
Toernooiveld, 6525 ED Nijmegen, the Netherlands

I am a member of a club which argues that one can get from high energy spectroscopy surprisingly good values for the (Hubbard) correlation energy of narrow band electrons, for the strength of hybridization energy of the narrow band states, and for the degrees of occupation of these levels. Working in this area can have its frustrations. On the one hand, if one is honest, one must admit that we are normally working with models in order to interpret the results. One thus wishes to discuss not only possible refinements of the models but also their limitations. On the other hand it is always difficult to live with the fact that many people do not seem to be aware about the possibilities of the high energy spectroscopies, or do not want to know. And if you talk about the problems in front of the people who do not know (or do not want to know) you do not make them very enthusiastic about what you are trying to do.

The Hubbard correlation energy, U_{eff} is normally described as the difference between the first ionization energy and the first electron affinity of a system. For instance if we have an array of d^9 (or f^7) atoms then U_{eff} is the energy required to change one atom to d^8 and a second to d^{10} (or one f^7 to f^6 and a second to f^8). The symbol U_{eff} is used to imply that it is an effective Coulomb correlation energy that we are talking about where all the valence electrons have been allowed to adjust their distribution in response to removal or addition of an electron. Y. Baer's group[1] made a large impact in the area of Narrow Band phenomena by putting together photoelectron (XPS) and Bremsstrahlung Isochromat (BIS) spectroscopy of the rare earths to obtain values of the ionization potentials and electron affinities with respect to the Fermi level (E_F) of the valence states, and hence U_{eff}. It should be noted that as the strength of hybridization of the localized states with the valence states is increased the spectra shape changes (see fig. 1) and new peaks may be observed near E_F. The exact details of the changes depend on the strength of the hybridization and the degeneracy of the levels involved.

A second method of measuring U_{eff}, using Auger spectroscopy, was pioneered by Cini and Sawatzky[2]. The idea is that the Auger transition in which a core hole is filled by a valence electron, with emission of a second valence electron ($\underline{C} \rightarrow \underline{VV} + e^-$) is a local transition. If the valence states are highly localized then the two final state holes will be on one site. The two hole binding energy is simply the binding energy of the core electron (which can be measured by XPS) minus the kinetic energy of the Auger elec-

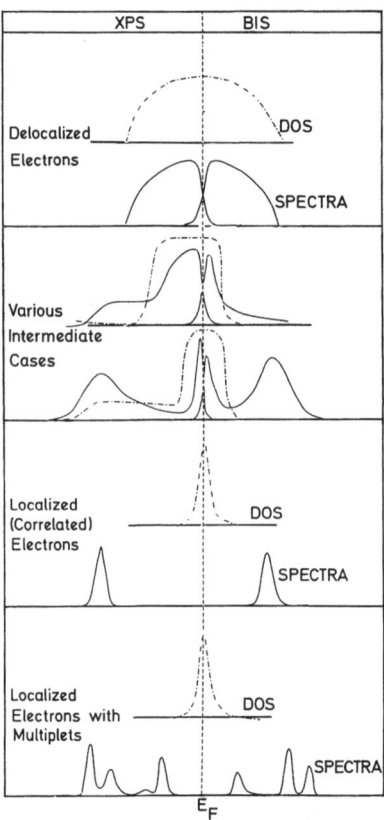

Fig. 1.   Schematic diagram of the development of the density of states
and XPS/BIS spectra with increasing localization (U/W) of valence
electrons.

tron. This may be compared with the sum of the binding energies of two
separate holes (measured by valence band photoemission). The difference is
U_{eff}. For instance for Cu, and La(Y)Ba(Sr)CuO compounds L_3VV Auger spectrum
shows the Cu d^8 energy to lie 11-13 eV above the ground state and photoemission shows that it costs on average about 2-3 eV to remove a single Cu
d electron[3]. The effective correlation energy here is that required to
localize two single Cu d holes on one site and that is of the order of
7 eV.

It should be added that if U_{eff} is not much greater than the strength
of hybridization, the spectral function takes on a more complicated form
(see fig. 2) and determination of U_{eff} is slightly less trivial[2,3].

The sort of questions one might discuss about the two experiment methods
for determining U_{eff} sketched above are:

1.   Are there really no serious objections to calling the quantities measured U_{eff}?
2.   Is U_{eff} really quite constant for a given state (e.g. 3d, 4f) in a given
element[3] regardless of chemical environment?
3.   The spectroscopic measurements often show evidence of atomic-like
multiplet splittings[3]. It is clear that the multiplet interactions
are not as well screened by the valence electrons of a solid as the

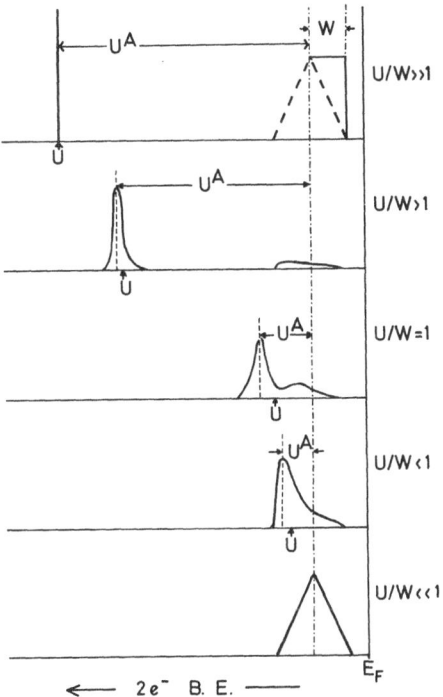

Fig. 2.   CVV spectra calculated for a rectangular DOS of width W. Spectra
           have been calculated with the Cini approximation [2,3] using several
           values of U/W (figure taken from ref. 3).

spherical Coulomb interaction. Does the observation of multiplets mean
that we should regard a material as having many different values of
U_{eff}? What is the consequence of the multiplet effects for the ground
state properties of the solid?

4.   The Cini-Sawatzky formalism for Auger lineshapes is only good for
      (nearly) filled bands. There is no satisfactory treatment for partially
      filled bands. Are experimental values form the earlier transition metal
      atoms misleading?

      There has been a lot of work done to try to extract estimates of the
strength of valence band hybridization from core level lineshapes of rare
earth materials. The models of Gunnarsson-Schönhammer (GS)[4] and Toyozawa-
Kotani[5] which are the basis of this method utilize the concept of a screening
orbital. When a core hole is created all the wave functions of the other
N-1 electrons of the systems relax to adjust to the new potential. In the
core-ionized state of lowest energy the valence electrons build up density
around an atom where a core hole has been created (as in XPS for example).
The distribution of this extra charge may look like that in an atomic orbital, so that one can speak of "screening orbitals". In the sudden approximation the valence electron distribution, averaged over all the possible
final states of a given core hole creation, is the same as in the ground
state.

      The G-S model is suitable for much solid state work because it in-
corporates partial occupation of the screening orbitals and hybridization
effects as well as the dynamics of photoemission. This is done by giving
the screening orbital a width and energy in both the initial and final
states, as in fig. 3, which illustrates the results of numerical calculations[4,6].

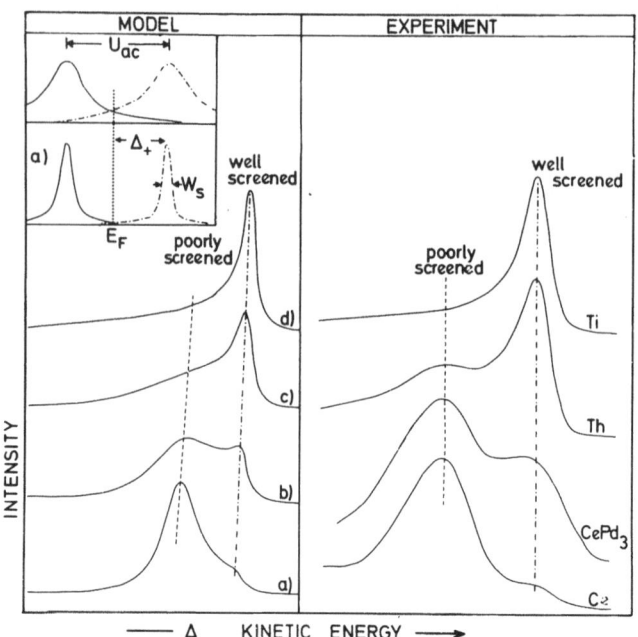

Fig. 3.   Left: illustration of the Schönhammer-Gunnarsson model of screening
and its implications for the core level line-shapes. The values
of $\Delta+/W$ were d=0.94, c=0.75, b=0.56, a=0.38 and U_{ac} was 1.5 $\Delta+^6$.
Right: Ce $3d_{5/2}$, Th $4f_{7/2}$, and Ti $2p_{3/2}$ XPS peaks from Ce, CePd$_3$,
Th and Ti. The peak binding energies are 883, 333 and 454 eV for
the Ce, Th and Ti levels respectively.

     In the initial state of photoemission the screening orbital is at an
energy $\Delta+$ above the Fermi level and has a width, W, which represents the
coupling to the other levels of the system. In the final state, in the
presence of a core hole, the screening level is pulled down U_{ac} with respect
to the Fermi level by the core hole potential. The level is now $\Delta-$ below
E_F. In this model the probability that the screening orbital will be occupied
after core level photoemission is related to the position of the orbital
and its coupling to the other orbitals. Thus if the screening orbital is
narrow and far above E_F in the initial state as shown in inset a, then its
occupancy is low. There is then very little overlap between the initial
state and the final state with the screening level occupied. This means
that the probability of a transition to the "well-screened" final state is
small. Most of the XPS intensity will be found in the "poorly screened"
peak lying approximately $\Delta-$ to higher BE and corresponding to the transition
to a final state with the screening level almost unoccupied as shown in
curve a. If, on the other hand, the width W of the screening level is of
the same order as its position $\Delta+$, the level is strongly coupled to the
system and its extensive tail below E_F can be interpreted as considerable
occupancy, even in the ground state. Conversely in the final state W is
also comparable to $\Delta-$ and there is considerable unoccupied character in
the well-screened final state. The probability that the screening orbital
becomes occupied in the final state of core-level photoemission, is then
high and most of the XPS intensity is found in so-called well-screened
peak, as shown in curve d).

     When we thought that we could see a link between what we actually
observed in XPS and the calculations of the G-S model, people were pestering

166

me to publish in Phys. Rev. Lett. because they said it was the only way to get to be a professor. In order to get the work into Phys. Rev. Lett. we inserted some sentences about how this wonderful effect could actually be used to measure the strength of coupling (hybridization) between the f levels and the valence states. The paper received the ultimate insult: acceptance for Phys. Rev. Lett. without comment. To our embarrassment a lot of people put in a lot of work to make the measure semi-quantitative, and the method has been used for upwards of 60 La and Ce compounds, see e.g. J.W. Allen's text. However, several things were underestimated in our first paper. In particular we did not realize how strongly the positions of the peaks can be influenced by the hybridization when this is larger than a few hundred meV so that for CeO_2 (see e.g. Kotani's text) the early actinides, and also for transition-metal compounds, more complicated calculations may be necessary. In those cases the mixing between the different final states can be large enough to make even the assignment of the XPS peaks (e.g. to f^1 and f^0) doubtful.

Nevertheless the basic idea of using core level spectral lineshapes to estimate ground state electronic structure parameters is finding increasing interest and not only XPS, but also XAS is widely used. The sort of questions that may still need some discussion are:

1. Some non-spectroscopists argue that these high energy scale spectroscopies are irrelevant for ground state and near ground state effects because the timescale of the experiment is different. Have their arguments really been answered?
2. Is it the timescale, or the size of the perturbation of the experiment which is relevant?
3. What do we know about the adequacy of the sudden approximation?
4. It is generally assumed that the hybridization potential mixing states with different "narrow band electron counts" is not affected by the number of core holes. Do we know anything about the validity of this assumption?
5. It has been stated that the hybridization of the f states is weaker in heavier rare earths. Is this true? How do the relative importances of degeneracy and hybridization potential change between La and Ce?
6. Mixing of the basis states of the final state produces a tendency to enhance the lowest energy state in any core level spectroscopy. How does this affect the attempts to determine valency from (for instance) Ce L_3 XAS without resort to model calculations?
7. How consistent are the values for strengths of hybridization of narrow bands taken from different high energy spectroscopies?
8. What are the prospects for core level studies of actinides, such as U, Pu, Am, etc.?

REFERENCES

1. P.A. Cox, J.K. Lang, and Y. Baer, J. Phys. F 11:113 (1981); 11:121 (1981).
2. M. Cini, Solid St. Comm. 24:681 (1977); and Phys. Rev. B15:2788 (1978); G.A. Sawatzky, Phys. Rev. B34:504 (1977).
3. P.A. Bennett, J.C. Fuggle, F.U. Hillebrecht, A. Lenselink and G.A. Sawatzky, Phys. Rev. B27:2194 (1983).
4. O. Gunnarsson and K. Schönhammer, Phys. Rev. Lett. 50:604 (1983); Phys. Rev. B28:4315 91983) and references therein.
5. A. Kotani and Y. Toyozawa, J. Phys. Soc. Japan 35:1073-1082 (1973).
6. J.C. Fuggle, M. Campagna, Z. Zołnierek, R.Lässer and A. Platau, Phys. Rev. Lett. 45:1597 (1980).

OBSERVATION OF LOW-ENERGY EXCITATIONS IN LANTHANIDE MATERIALS

BY HIGH-RESOLUTION PHOTOEMISSION

W.-D. Schneider and Y. Baer

Institut de Physique, Université de Neuchâtel

CH-2000 Neuchâtel, Switzerland

INTRODUCTION

One of the fundamental motivations for electron spectroscopic studies is to deduce ground state properties of electronic systems from their excitation spectra. Among the high-energy spectroscopies photoemission has become very popular because it is commonly considered as one of the most direct methods to probe the density of the electronic states (DOS) in solids. However, when the correlation among the band electrons increases, the relationship between measured spectra and DOS is no longer straightforward[1]. In the extreme case of the 4f-states, which are sufficiently localized to participate only very weakly to bands, the simple single-particle picture becomes almost meaningless and fails to account for many properties involving their excitations. In particular, at low temperatures, the quasi-equilibrium methods (specific heat, susceptibility, transport properties) applied to the study of systems containing lanthanide elements, reveal very unusual manifestations, like Kondo-, mixed-valence, or heavy-Fermion behaviour[2]. The single-impurity model of Anderson, allowing the mixing of one localized 4f-state with extended states, has proven to yield a first approach to important aspects of this problem [3,4,5]. It has been used to calculate the spectral functions of high-energy spectroscopies. The comparison with experiments, performed with a resolution of about two orders of magnitude larger than the characteristic energy scale of these many-body manifestations, has demonstrated that this model contains the essential mechanisms allowing one to account for the observed final-state excitations [6-11]. In this way the parameters defined in the Anderson-model can be determined and, within the same model, the low-energy thermodynamic properties can be calculated [12]. However, it is an experimental challenge to observe directly the low-lying many-body excitations close to the ground state and responsible for the unconventional low-temperature properties of hybridized f-systems. Unfortunately, with the commonly achieved experimental resolution this goal remained unrealistic for a long time. We have shown [13-18] that already with a resolution improvement of a factor of 10 (< 20 meV) these narrow bands of low-lying excitations are accessible to experiment and that the information extracted from these high-resolution photoemission spectra can be linked to results obtained by low-energy methods. The aim of the present paper is to demonstrate that this approach reveals original aspects of condensed matter physics.

CRYSTALLINE ELECTRIC FIELD AND DEGENERACY OF THE GROUND STATE

An important issue for the low-temperature properties of hybridized f-systems, which has been recognized and taken into account only recently in the analysis of electron spectroscopic data, is the influence of the crystalline electric field (CEF). Depending on the crystal symmetry, the degeneracy $N_f = 6$ of the $4f_{5/2}^1$ spin-orbit component of Ce is lowered to 4 or 2 in the lowest crystal state. Consequently a Gunnarsson-Schönhammer-type many-body calculation[3], when performed only to the lowest order in $1/N_f$, certainly does not yield a well-converged ground-state. Moreover, the low-energy parameter δ representing the energy lowering of the ground state due to hybridization, is reduced with decreasing degeneracy[19]. It has been confirmed that the low-energy parameter δ, extracted from this refined spectra analysis, is consistent with the analysis of the specific heat [16,20]. When the splitting Δ_{CEF} between the two lowest crystal-field levels exceeds δ, these levels become apparent in sufficiently well resolved spectra[20] opening the possibility to use high-resolution photoemission as a complementary tool to inelastic neutron and light scattering in the study of CEF levels.

ENERGY SCALE AND TEMPERATURE RANGE FOR THE SPECTROSCOPIC OBSERVATION OF THE KONDO RESONANCE

The analysis of the low-lying excitations, observed with photoemission shows that already the single-impurity model at $T = 0$ contains the relevant mechanisms allowing us to simulate the spectra. Most recently, this type of spectra analysis has been extended to finite temperatures, enabling a detailed study of the high-temperature collapse of the Kondo-resonance in CeSi$_2$ [20]. This compound has a linear coefficient of the specific heat $\gamma(T \rightarrow 0) \approx 100$ mJ/mole K^2, so that from the relationship $\gamma \sim 1/\delta$ a favorable situation for studying the temperature dependent behaviour of the low-lying excitations was expected. In Fig.1 (a)-(d) the f-contribution to the experimental spectra is shown in a narrow energy range above E_F, displaying essentially the weakly hybridized $4f_{7/2}$ (around 280 meV) and $4f_{5/2}$ (around E_F) final states. These data are obtained from differences of HeII and HeI spectra, weighted by the same cross-section ratio for all temperatures. With increasing temperature the first peak above E_F is progressively washed out and has completely disappeared at 300 K where the thermal broadening of the Fermi edge is about 100 meV ($\sim 4k_BT$). The $4f_{7/2}^1$ final states around 280 meV form a rather symmetric peak at 15 K, which is also attenuated with increasing temperature but remains clearly visible. The experimentally derived f-spectra (a)-(d) are compared with the f-spectral functions (e)-(h), computed for the corresponding sample temperatures and broadened by the instrumental resolution of 18 meV. The calculation has been performed within the Anderson single-impurity model using the Slave Boson approach[21] in the non-crossing approximation [22] and including spin-orbit and crystal field splittings. The tetragonal symmetry of CeSi$_2$ lowers the degeneracy N_f of the ground state from 6 to 2. The CEF splitting between the ground state and the first excited doublet was assumed to the 35 meV, which is compatible with the experimental spectra. With the single-impurity parameters $\epsilon_f = -1.6$ eV and $\Delta = 95$ meV a ground state f-occupation $n_f = 0.97$ and a low-energy parameter $\delta = 3$ meV corresponding to a Kondo temperature of $\delta/k_B = 35$ K are obtained from the computation. Apart from the width of the experimental structures which is partly determined by lifetime broadening not included in the calculation, the agreement between experiment and model for all temperatures is striking. It proves that this many-body calculation yields a correct description of the temperature dependence of our experimental spectra. The formation of the singlet ground state giving rise to the Kondo resonance close to E_F requires a sharp cut-off in the occupation of the extended states and therefore can only develop at sufficiently low temperatures ($T \stackrel{<}{\sim} T_K$). When $k_B T$ increases and exceeds $k_B T_K$, non-singlet states of higher energy become more and more populated so that the Fermi liquid behaviour transforms gradually into an atomic-like behaviour. The spectra of Fig. 1 (a)-(d) yield the first unmistakable observation with photoemission of this progressive transition appearing as a collapse of the extremely narrow and intense distribution of the low-energy many-body excitations when the temperature is raised above T_K.

The uppermost curve (i) in Fig. 1 represents the calculated f-spectrum at T = 15 K without instrumental broadening. It shows that with still improved resolution the crystal field excitation at 35 meV could be clearly discriminated from the Kondo resonance, pinned at E_F. It remains that an important fraction of the experimentally observed intensity around E_F must be attributed to this many-body resonance.

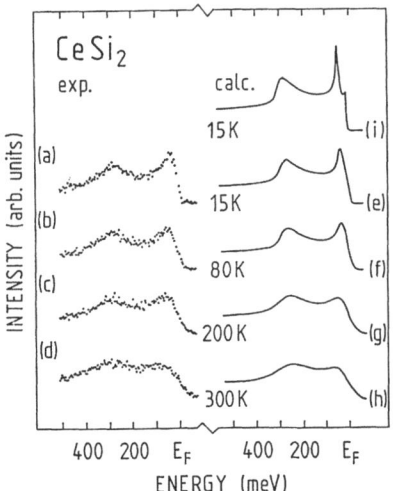

Fig. 1. Comparison between experimentally derived and computed f-spectra of $CeSi_2$. The curves (a)-(d) represent differences of weighted spectra, taken at HeII and HeI excitations energies. The curves (e)-(h) are the result of a many-body calculation for the corresponding temperatures including $4f^1$ spin-orbit and $4f^1_{5/2}$ crystal field splittings. A convolution with Gaussians of 18 meV FWHM has been performed in order to account for the finite instrumental resolution. Curve (i) represents the calculated f-spectrum for T = 15 K without instrumental broadening (Ref. 20).

THE DIFFERENT REGIMES OF THE SINGLE-IMPURITY MODEL

In Ce-systems, the many-body parameters deduced from the spectroscopic analysis, are always characteristic for the Kondo-regime ($\varepsilon_f \gg \Delta$, $n_f \approx 1$)[12]. In heavier rare earths, the single-impurity parameters are likely to become different and other regimes are conceivable [21-23]. This is the case for $YbAl_3$, which has been found to belong to the mixed valence regime ($\varepsilon_{f_{13}} \approx \Delta$, $n_{f_{14}} \approx 13.5$)[24], where the ground state consists of a mixing of $4f^{13}$ and $4f^{14}$ wavefunctions with approximatively equal weight.

All these recent data have provided new evidence that the single-impurity model is well suited to describe the fundamental mechanisms governing the 4f-conduction band hybridization in rare earth materials.

OUTLOOK

High-resolution photoelectron spectroscopy has proven to be well adapted to probe the low-lying excitations closely related to the ground state. So far the ultimate resolution of these methods seems to be only determined by the lifetimes of the states which are involved but is not affected by the high-energy of the induced transitions [10,25]. For the extreme conditions of heavy-Fermion systems with γ-values of ~ 1 J/mole K^2, corresponding to δ-values of < 1 meV, very small spectral weights $n_f(n_f-1) < 0.01$ of the Kondo resonance are expected [14]. At very low temperatures coherence effects are believed to induce a splitting of the Kondo peak[26] at an energy scale $(1/N_f)\delta$[25]. A test of these predictions with photoemission techniques, carried out at the required low temperatures, remains an experimental challenge for the future.

In order to obtain a complementary picture of the Kondo resonance in electron excitation spectra, the development of high-resolution inverse photoemission is highly desirable. For example in Ce-systems, where the relative intensity of the Kondo peak probed by electron addition is about 13 times stronger than in electron subtraction (photoemission), high-resolution will provide the possibility to observe the low-energy excitations and the predicted spin-orbit side band [4], separated by 0.28 eV. Up to now the available instrumental resolution of 0.6 eV allows us only to observe a superposition of these features. [27,18].

It must be emphasized that in addition to ε_f the hybridization strength Δ is the main adjustable parameter in the model, so that methods determining Δ from first principle band structure calculations are very valuable[28]. For example, in view of the 4f-wave function contraction across the lanthanide series, the surprising fact that the hybridization strength in Yb-materials is of the same order of magnitude as in Ce-materials[24], deserves further attention.

The 4f photoemission spectra of the Lanthanides can not be simply interpreted in terms of single-particle DOS but they reflect the pronounced many-body nature of their electronic structure. At low energy the excitations form a spectrum showing similarities with an extremely narrow metallic band, whereas at higher energy they have a localized character reminiscent of purely atomic final states. This duality, also observed as a function of the temperature in other properties, is a striking but not completely elucidated consequence of a weak hybridization.

ACKNOWLEDGEMENTS

We wish to thank F. Patthey , J.-M. Imer and B. Delley for a fruitful collaboration. This work has been supported by the Swiss National Science Foundation.

REFERENCES

1. G. Treglia, F. Ducastelle, and D. Spanjaard, J. Phys. (Paris) 41, 281 (1980); Phys. Rev. B21, 3729 (1980), and references therein.

2. G. R. Stewart, Rev. Mod. Phys. 56, 755 (1984).

3. O. Gunnarsson and K. Schönhammer, Phys. Rev. Lett. 50, 604 (1983); Phys. Rev. B28, 4315 (1983).

4. N. E. Bickers, D. L. Cox and J. W. Wilkins, Phys. Rev. Lett. 54, 230 (1985).

5. P. A. Lee, T. M. Rice, J. W. Serene, L. J. Sham, and J. W. Wilkins, Comments Cond. Mat. Phys. 12, 99 (1986).

6. J. C. Fuggle, F. U. Hillebrecht, J.-M. Esteva, R. E. Karnatak, O. Gunnarsson, and K. Schönhammer, Phys. Rev. B27, 4637 (1983).

7. J. C. Fuggle, F. U. Hillebrecht, Z. Zolnierek, R. Lässer, Ch. Freiburg, O. Gunnarsson and K. Schönhammer, Phys. Rev. B27, 7330 (1983).

8. F. U. Hillebrecht, J. C. Fuggle, G. A. Sawatzky, M. Campagna, O. Gunnarsson and K. Schönhammer, Phys. Rev. B30, 1977 (1984).

9. W.-D. Schneider, B. Delley, E. Wuilloud, J.-M. Imer, and Y. Baer, Phys. Rev. B32, 6819 (1985).

10. Y. Baer and W.-D. Schneider, in : "Handbook on the Physics and Chemistry of Rare Earths, K. A. Gscheidner, L. Eyring and S. Hüfner, eds. (North-Holland, Amsterdam 1987) Vol. 10, chap. 62.

11. J.-M. Imer and E. Wuilloud, Z. Phys. B66, 153 (1987).

12. J. W. Allen, S. J. Oh, O. Gunnarsson, K. Schönhammer, M. B. Maple, M. S. Torikachvili, and I. Lindau, Advances in Physics 35, 275 (1986), and references therein.

13. F. Patthey, B. Delley, W.-D. Schneider, and Y. Baer, Phys. Rev. Lett. 55, 1518 (1985); ibid 57, 270 (1986); ibid 58 1283 (1987) (E).

14. F. Patthey, W.-D. Schneider, Y. Baer, and B. Delley, Phys. Rev. B34, 2967 (1986).

15. F. Patthey, S. Cattarinussi, W.-D. Schneider, Y. Baer, and B. Delley, Europhys. Lett. 2, 883 (1986).

16. F. Patthey, W.-D. Schneider, Y. Baer, and B. Delley, Phys. Rev. B35, 5903 (1987).

17. Y. Baer, F. Patthey, W.-D. Schneider, B. Delley, J. Mag. Magn. Mat. 63&64, 503 (1987).

18. W.-D. Schneider, F. Patthey, Y. Baer, B. Delley, in : "Giant Resonances in Atoms, Molecules and Solids, NATO ASI Series B, eds. J.-P. Connerade, J.-M. Esteva, and R. E. Karnatak (Plenum, New-York, 1987) p. 461.

19. A. C. Hewson, D. M. Newns, J. W. Rasul, N. Read, H. U. Desgranges, P. Strange, in : Theory of Heavy-Fermions and Valence Fluctuations, eds. T. Kasuya and T. Saso, Springer Series in Solid State Sciences, Vol. 62 (Springer, Berlin 1985) p. 134; O. Gunnarsson and K. Schönhammer, ibid. p. 100.

20. F. Patthey, W.-D. Schneider, Y. Baer, B. Delley, to be published

21. P. Coleman, Phys. Rev. B29, 3035 (1984).

22. Y. Kuramoto, Z. Phys. B53, 37 (1983).

23. H. Kojima, Y. Kuramoto, and M. Tachiki, Z. Phys. B54, 293 (1984).

24. F. Patthey, J.-M. Imer, W.-D. Schneider, Y. Baer, B. Delley and F. Hulliger, to be published.

25. O. Gunnarsson and K. Schönhammer, in "Handbook on the Physics and Chemistry of Rare Earth", K. A. Gscheidner, L. Eyring and S. Hüfner, eds. (North-Holland, Amsterdam 1987) Vol. 10, chap. 64.

26. N. Grewe, Solid State Commun. 50, 19 (1984).

27. E. Wuilloud, H.R. Moser, W.-D. Schneider and Y. Baer, Phys. Rev. B28, 7354 (1983).

28. R. Monnier, L. Degiorgi and D. D. Koelling, Phys. Rev. Lett. 56, 2744 (1986).

THE IMPORTANCE OF ATOMIC RELAXATION IN CORE-LEVEL SPECTRA

FOR NARROW-BAND SYSTEMS

Nils Mårtensson

Institute of Physics

Box 530, S-751 21 Uppsala, Sweden

In recent years various high energy spectroscopies, such as photo-electron spectroscopy, Auger electron spectroscopy, inverse photoemission and X-ray absorption spectroscopy, have proven to be most useful in probing the electronic structure of narrow-band systems.[1] Much of the theoretical work in this connection has started out from the utilization of the impurity Anderson Hamiltonian. One important question in the description of the results from these different techniques is to what extent the same parameters should be obtained for the different experimental situations. For instance, to what extent are the parameters used to describe the spectral shapes modified due to the presence of the core-hole?

In this connection it is of interest to consider some properties of core level spectra from free atoms. The most clear-cut situation is obtained for closed-shell atoms. The experimental situation is particularly favourable for the rare gases. Figure 1 shows the total 1s core level spectrum from Ne including the various shake-up (single shake-up, double shake-up, etc.) and shake-off contributions.[2] The inelastic scattering contribution to the spectrum is very low and furthermore the small remaining inelastic contri-bution can be rather accurately corrected for. This makes it possible to determine the total intensity of the combined shake-up and shake-off satel-lite spectrum with good accuracy. A total satellite intensity of 37(4)% relative to the main line is obtained in this way for the Ne 1s photoelectron spectrum. The most prominent individual shake-up peaks have intensities of 3.15 and 2.85%.

When we are using various types of satellite spectra for the analysis of the electronic structure of solids it is important to recognize that there could also be a substantial satellite contribution which has a purely atomic origin as in Ne. It is of course impossible to make a clear separation between the atomic and the solid state contributions to the satellite spectra. However, it is clear that in order to describe the finer details of a shake-up spectrum in a solid the theoretical description has to include those aspects of the electron charge distribution which give rise to the satellites also in the atomic case.

As in extended systems, the satellites in free atoms are due to the relaxation of the electron cloud around the core hole. In a solid a large fraction of this relaxation is due to the tendency of the conduction elec-trons to neutralize the core-ionized site. For a free atom the relaxation is mainly due to a contraction of the orbitals in the core ionized atom. Theoretical calculations of high energy excited shake-up spectra are usually treated within the so-called shake-model.[3] the most sophisticated version of this model incorporates multiconfiguration theory. In the case of a fast outgoing photoelectron the shake-theory implies that the intensities of the shake-up satellites can be calculated by only considering initial and final state wave-functions; the variation of the transition moment with energy is usually neglected. The degree of sophistication of different calculations spans from simple calculations treating only overlap between the two orbitals involved in the shake-up process to calculations taking full account of the initial state correlation (ISCI) as well as of final state correlation (FISCI).

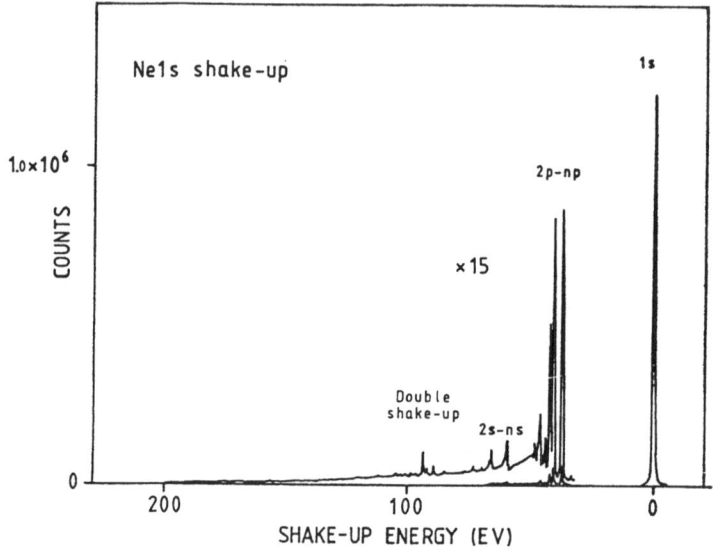

Fig. 1.   Total 1s core level XPS spectrum of Ne.

An impression of how the relaxation influences the shake-up intensities can be obtained in a most simplified way. In the case of the 2p-np shake-up processes in Ne the transition moments for the various final states will to a first approximation be proportional to the overlap between the wave functions for the $1s2s^2 2p^6$ and $1s2s^2 2p^5 np$ configurations. If we simply treat these wave functions as product wave functions of the individual orbitals, the difference in magnitude of these matrix elements will be dominated by the terms $\langle 2p|np'\rangle$ the prime is used to denote the final state wave-functions). If there is no relaxation, the initial and final state wave functions are identical. This implies in particular that the 2p and np' orbitals are orthogonal and the shake-up transition moments are zero. In this case all intensity will be in the $1s2s^2 2p^6$ final state line (the main line). On the other hand, with increasing modification of the final state orbitals, spectral weight can be expected to be transferred to the shake-up states.

With this picture in mind it is interesting to investigate what happens
e.g. to the 4f shell in the early lanthanides upon core-ionization. As a
simple measure of this effect the expectation value of r can be used. The
radial expectation values <r> have been calculated using a Dirac-Slater
program. For Ce a contraction of the 4f orbital of more than 10% is obtained
in these calculations. In order to get a feeling for what is normal it can
be noted that the corresponding contraction of the 2p orbital in ne is
only half as big. In this situation it is thus reasonable to assume that
such "atomic" effects can have a significant influence also on Ce core-
level spectra. This implies that the neglect of these effects could lead
to inconsistent results when information about the solid state electronic
structure is derived from different types of spectroscopies.

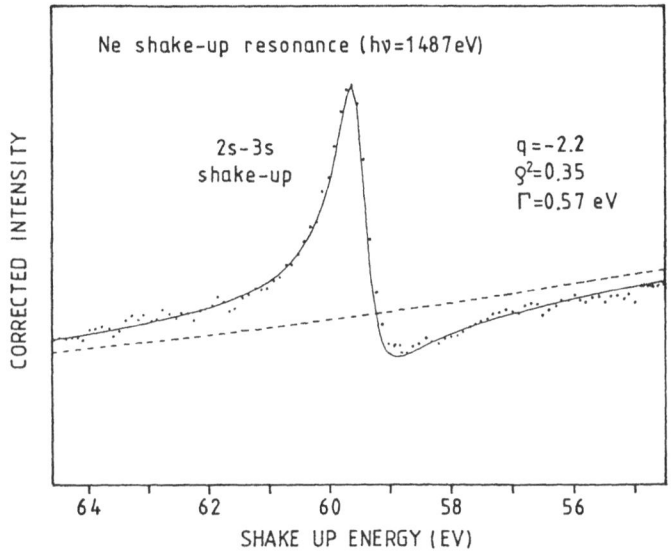

Fig. 2.   2s-3s shake up resonance in Ne.

In the discussion above the terms shake-up and shake-off have been
used to denote processes in which the photoionization is accompanied by
the simultaneous excitation of an electron to a discrete state (shake-up)
and to a continuum state (shake-off). The Ne spectra clearly show that it
is important to consider also the interaction between these two types of
channels. The shake-up states which lie above the shake-off thresholds are
found to have Fano-like line shapes.[4] This is clearly seen from Fig. 2
which shows the 2s-3s shake-up resonance in Ne. The satellite line has a
clear Fano-type line-shape due to the interaction with the 2p-ϵp shake-off
continua with the same symmetry. This interaction leads to shifts, broaden-
ings and modifications of the intensities of the satellite lines. These
effects will occur for all shake-up states which lie above the first shake-
off threshold. However, usually the investigated shake-up structures are

too broad to allow a direct detection of this kind of effect even if it is influencing the spectrum considerably.

To summarize, this contribution has been meant as a reminder that there might be substantial "atomic" effects in solid phase core level satellite spectra. This should be considered when inconsistencies are found between the results from different spectroscopies. If necessary these effects could certainly be included in the treatment of core-level spectra. However, it is still unclear how large these effects will be in various narrow-band situations.

REFERENCES

1. J.W. Allen, S.-J. Oh, O. Gunnarsson, K. Schönhammer, M.B. Maple, M.S. Torikachvili, and I. Lindau, Adv. in Physics 35:275 (1986).
2. N. Mårtensson, S. Svensson, and U. Gelius, to be published.
3. R.L. Martin and D.A. Shirley, Phys. Rev. A13:1475 (1976).
4. S. Svensson, N. Martensson, and U. Gelius, to be published.

FINAL STATE INTERACTION IN L_3 ABSORPTION SPECTRA OF Ce AND La COMPOUNDS

A. Kotani

Department of Physics
Tohoku University
Sendai 980, Japan

Core level spectroscopy is one of the most powerful tools in the study of 4f electron state of rare earth systems. Spectra of 3d core photoemission (3d-XPS) and 2p core photoabsorption (L_3-XAS) have been used most extensively in experimental observations. In the final state of these spectra, a core hole is left behind, and the attractive core hole potential $-U_{fc}$ to the 4f state gives a strong influence on the spectral shape, as first pointed out by Toyozawa and the present author.[1] Therefore, in order to obtain the information on the 4f state it is necessary to analyze the experimental data quantitatively by taking account of the final state interaction. For 3d-XPS, such a quantitative analysis was first made by Gunnarsson and Schönhammer[2] with the use of the impurity Anderson model, and their method has been applied successfully to the analysis of 3d-XPS in various materials, e.g., metallic and insulating systems including Ce and La. For L_3-XAS, however, detailed mechanism of the final state interaction has hardly been studied theoretically until very recently. It is the purpose of this note to discuss the mechanism of L_3-XAS in Ce and La compounds.

According to experimental data, characteristic features of L_3-XAS in Ce and La compounds are usually different from those of 3d-XPS. The difference between 3d-XPS and L_3-XAS is observed most clearly in the insulating Ce compound, CeO_2. The 3d-XPS of CeO_2 exhibits three peaks, aside from the spin-orbit splitting of 3d level. The energy separation between the outermost peaks is about 16 eV. On the other hand, the L_3-XAS has only two peaks, whose energy separation is about 8 eV. This result suggests that 4f configurations in final states of 3d-XPS and L_3-XAS are different. As an origin of this difference, there may be two possibilities: (i) A difference in the core hole potential $-U_{fc}$ between 2p and 3d core holes. (ii) The existence of a photoexcited 5d electron in the final state of L_3-XAS.

Jo and the present author[3] considered that the point (ii) is essentially important, and proposed a mechanism by which the 3d-XPS and L_3-XAS are explained consistently. Let us consider the impurity Anderson model with a filled valence band. The system is composed of a 4f level, a filled valence band (O 2p band), an empty conduction band (Ce 5d band), and a core state. The hybridization V between the 4f and valence band states and the Coulomb interaction U_{ff} between 4f electrons are taken into account. In the final state of 3d-XPS, a 3d core electron is removed and the core hole potential $-U_{fc}$ acts on the 4f state. In the final state of L_3-XAS, on the other hand, a 2p core electron is excited to the 5d band, and we take account of interactions U_{fd} between 5d and 4f electrons and $-U_{dc}$ between the 5d electron and the core hole. It is assumed that the core hole potential $-U_{fc}$ is the same for the 3d and 2p core holes (disregard of the possibility (i)). In Fig. 1, the calculated L_3-XAS and 3d-XPS (inset) are shown. The parameter values V, U_{ff}, U_{fc} etc. are determined so as to reproduce the experimental three-peak structure of 3d-XPS, and then the average 4f electron number in the ground state is found to be about 0.5. In this sense CeO_2 is in the "mixed valence state". When U_{fd} and U_{dc} are disregarded, L_3-XAS has three peaks as shown with the dashed curve, because it is nothing but a broadened version of 3d-XPS. When U_{fd} and U_{dc} are taken to be 4 eV and 5 eV, respectively, L_3-XAS has two peaks in agreement with experimental data. The role of the attractive potential $-U_{dc}$ is to localize the excited 5d electron near the core hole site. Then, a reconstruction of 4f configurations occurs by the effect of U_{fd}, and it results in the appearance of two peaks corresponding mainly to $4f^0$ and $4f^1$ final states. Recently, very similar experimental data of 3d-XPS and L_3-XAS have also been observed in CeF_4. Furthermore, the importance of U_{fd} and $-U_{dc}$ has been confirmed in the analysis of L_3-XAS in insulating La compounds, La_2O_3 and LaF_3.[4]

On the possibility of (i), a theoretical calculation by Herbst and Wilkins[5] indicates that U_{fc} of 3d core hole in Ce metal is larger than that of the 2p core hole by 1.3 eV, but no calculation has been made for CeO_2. According to recent 3p core spectroscopies in CeO_2, it is confirmed that the point (i) is less important than (ii). The 3p-XAS of CeO_2 observed by Kaindl et al.[6] is very similar to L_3-XAS, although the spectral width is larger. Therefore, the value of U_{fc} and U_{dc} are almost the same for 2p and 3p core holes. Furthermore, the 3p-XPS, which has very recently been observed by Bianconi,[7] can also be reproduced fairly well from the 3d-XPS of Fig. 1 by increasing the spectral width and by adding a background contribution. From these facts we can conclude that the value of U_{fc} is almost the same for 2p, 3p and 3d core holes. The difference in the experimental spectra between 3p-XAS and 3p-XPS can be explained only by the effect of U_{fd} and $-U_{dc}$. Because of the large width of the 3p core spectra, partly due to the intrinsic short lifetime of the 3p core hole and partly due to extrinsic experimental conditions, there

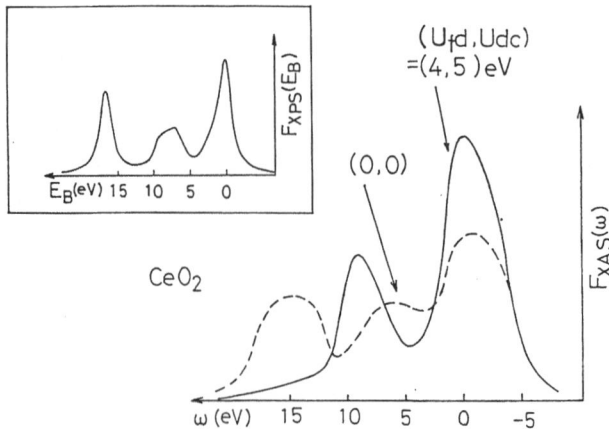

Fig. 1. Calculated results of L_3-XAS and 3d-XPS (inset) of CeO_2. Origins of the photon energy ω and the binding energy E_B are taken arbitrarily.

remains a possibility of the difference in U_{fc} between 3p and 3d (or 2p) core holes by about 1 eV, but it is too small to explain the difference between 3d(or 3p)-XPS and 2p(or 3p)-XAS. More precise experimental observations of 3p core spectra, and also 2p-XPS if possible, are highly desirable.

Finally we give some remarks on metallic systems. The present filled band Anderson model may be applicable to the analysis of metallic systems within the lowest order approximation of $1/N_f$ expansion (N_f being the degeneracy of 4f level), although the screening effect by conduction electrons is not taken into account explicitly. A rather preliminary analysis[4] of L_3-XAS in $LaPd_3$ shows that U_{fd} and U_{dc}, as well as U_{fc} and U_{ff}, are considerably smaller than those of insulating systems, but they are still important to explain the difference between 3d-XPS and L_3-XAS. Since the differnce between 3d-XPS and L_3-XAS itself is samller in metallic systems than in insulating ones, it may also be explained to some extent by the difference of U_{fc} between 2p and 3d core holes. It is necessary to calculate 3d-XPS and L_3-XAS by taking account of metallic screening effect explicitly. On the other hand, it is also necessary to comfirm experimentally the difference of U_{fc}. To this end, we hope that experimental observations of 3p-XPS and 2p-XPS in various metallic systems will be performed. We also propose that the X-ray emission spectrum due to the electronic transition from 3d to 2p levels will give an important information on the difference of U_{fc}.

References

1. A. Kotani and Y. Toyozawa, J. Phys. Soc Jpn. 37: 912 (1974).
2. O. Gunnarsson and K. Schönhammer, Phys. Rev. B28: 4315 (1983).
3. T. Jo and A. Kotani, Solid State Commun. 54: 451 (1985); A. Kotani and
   T. Jo, Proc. 4th Int. Conf. on EXAFS and Near Edge Structure,
   Fontevraud, 1986.
4. A. Kotani, M. Okada, T. Jo, A. Bianconi. A. Marcelli and J. C.
   Parlebas, J. Phys. Soc. Jpn. 56: 798 (1987).
5. J. F. Herbst and J. W. Wilkins, Phys. Rev. B26: 1689 (1982).
6. G. Kaindl, G. Kalkowski, W. D. Brewer, E. V. Sampathkumaran, F.
   Holtzberg and A. Schach v. Wittenau, J. Magn. Magn. Mater. 47-48:
   181 (1985).
7. A. Bianconi, unpublished.

THE INFLUENCE OF ELECTRONIC POLARIZATION ON THE

CORE XPS SPECTRA OF Ce-COMPOUNDS

W. Folkerts and C. Haas

Laboratory of Inorganic Chemistry, Materials Science Center
University of Groningen, the Netherlands

It is difficult to describe correctly the electronic structure of a system in which correlation of localized electrons and hybridization of these electrons with extended states both are important. The Anderson impurity model has been used in several forms to give insight in this problem [1,2]. However, in the Anderson Hamiltonian the screening of f- and core electrons by polarization is neglected. We show in this contribution that in some cases the approximation can lead to erroneous results for the value of the f-occupancy in the ground state $\langle n_f \rangle$, deduced from photoemission data.

Gunnarson and Schönhammer worked out a method for calculating photo-emission spectra of Ce compounds using the Anderson impurity model in the limit of large degeneracy of the f level[3]. Several authors[4,5] used their work to determine the parameters of the model from experimental spectra. These parameters are the energy of the f level ε_f with respect to the Fermi level, the effective Coulomb interaction between f electrons on the impurity U, the Coulomb interaction between an f electron and a core hole Q, and a hybridization parameter Δ, describing the coupling of the f-level with the band states. It is commonly realized that these parameters are effective quantities for which a precise interpretation is difficult. This is because several degrees of freedom of the system are not treated explicitly in the model, but rather in an effective manner. Electronic polarization of the host, due to terms of the form

$$\langle k\sigma(1)\, m\sigma'(2)\, |e^2/r_{12}|\, k'\sigma(1)m\sigma'(2)\rangle\, c^+_{k\sigma}c_{k'\sigma}n_{m\sigma'}$$

(k' and k refer to occupied and unoccupied band states, respectively, m to an electron localized on the impurity, and σ to the spin) causes changes of U, ε_f and Q with respect to the unscreened values U^*, ε_f^* and Q^* [6,7]

$$U = U^* - 2E_p$$

$$\varepsilon_f = \varepsilon_f^* - E_p (1 - 2n_o) \tag{1}$$

$$Q = Q^* - 2E_p$$

Here $(-E_p)$ is the polarization energy induced by a point charge $\pm e$ at the impurity site, and n_o is the number of valence electrons of the neutral impurity. Thus the effective parameter U respresents the on-site Coulomb interaction reduced by screening. The reduction $\Delta U_{scr} = U^* - U$ is quite large (for Ce $U^* \simeq 15$ eV [8], $U \simeq 7$ eV [5]), indicating that the polarization energy is appreciable. A large local screening cannot be caused by low energy excitations only, and configuration interaction involving high energy excitations is needed.

When constructing the wavefunctions of the system in the ground state and (XPS) final states, one has to take into account the polarization cloud. The propagation of the f electron through the lattice will be accompanied by the simultaneous motion of the screening cloud. This will lead to an enhanced effective mass (electronic polaron)[6,9]. The presence of a polarization cloud also influences the intensities of the peaks in the photoelectron spectra, from which the f-occupancy in the ground state $\langle n_f \rangle$ is deduced. This value of $\langle n_f \rangle$ is important in understanding low energy scale properties of the compound. In practice $\langle n_f \rangle$ is determined from core XPS spectra, as in valence XPS and BIS the intensities of peaks are often obscured by the presence of the ligand valence band and the Ce 5d band.

Schönhammer and Gunnarsson[10] discussed polarization effects for several cases, but did not consider cases in which a core hole is present. For two reasons we expect the polarization to be important in core XPS. First the polarization in initial and final states is very different, as a result of the core hole with charge +e. Second, the energy differences between final states in core XPS are large ($2\varepsilon_f - 2Q + U$ in zeroth order). This is of the same order of magnitude as the energy B of the excitations responsible for the polarization. These excitations may be electron-hole pairs, plasmons, excitons, etc. In metals a measure for B is the plasmon energy, in insulators the exciton energy[7,10].

First, we qualitatively discuss two limiting cases. If $E_p \gg \Delta$, the ground state can be written as

$$\Psi^O = \sum_n c^O_n |f^n\rangle \, \psi_{pol}(n) \tag{2}$$

and the final states with a core hole $\underline{c}$ as

$$\Psi^i = \sum_n c^i_n |f^n \underline{c}\rangle \, \psi_{pol}(n-1). \tag{3}$$

The wavefunction of the polarization cloud ψ_{pol} depends on the charge of the impurity. The intensity of the peaks in the XPS spectrum is given by

$$I^i = |\sum_n c^O_n c^i_n \, \langle \psi_{pol}(n) | \psi_{pol}(n-1) \rangle|^2 \tag{4}$$

Using a simple model for the polarization cloud we find that the overlap is given by

$$\langle \psi_{pol}(n) | \psi_{pol}(n-1) \rangle^2 = \exp(-2E_p/B) \tag{5}$$

which is independent of n. Therefore in this case ($E_p \gg \Delta$) the intensity of all peaks is reduced by the same factor with respect to the spectrum for $E_p = 0$.

If $E_p \ll \Delta$ the ground state is

$$\Psi^O = [\sum_n c^O_n |f^n\rangle] \, \psi_{pol}(\langle n_f \rangle^O) \tag{6}$$

and the final states are

$$\Psi^i = [\sum_n c^i_n |f^n \underline{c}\rangle] \, \psi_{pol}(\langle n_f \rangle^i - 1) \tag{7}$$

with

$$\langle n_f \rangle^i = \sum_n n |c^i_n|^2.$$

In this case the reduction factor of the intensity of peak i

$$\exp \; -(2E_p/B)(\langle n_f \rangle^0 - \langle n_f \rangle^i + 1) \qquad (8)$$

depends on i, and different relative intensities are found, as compared to the case for $E_p = 0$. The $|f^0\underline{c}\rangle$ like peak in the spectrum is affected most strongly by this polarization effect.

Usually we have to deal with intermediate cases, where polarization and hybridization are of the same order of magnitude. To study this region of parameters we use a simple model as discussed by de Boer et al.[7]. The Hamiltonian is

$$H = H_o + H_{pol}$$

$$H_o = \sum_{l=o}^{m} \{1[\varepsilon_f - Qcc^+ + \tfrac{1}{2}(1-1)U] \; g^+_1 g_1 +$$

$$+ \; T\sqrt{1(m+1-1)}(g^+_1 g_{1-1} + g^+_{1-1} g_1)\} + \varepsilon_c n_c \qquad (9)$$

$$H_{pol} = \sum_{k=o}^{n} \{kBv^+_k v_k + P\sqrt{k(n+1-k)}(cc^+ - \sum_{l=o}^{m} 1g^+_1 g_1)(v^+_k v_{k-1} + v^+_{k-1} v_k)\}$$

where g^+_1 is a creation operator for a many-electron unhybridized state with l f-electrons, and v_k^+ creates k excitons (or plasmons). The transfer T provides the hybridization of the f electrons with the band states. The degeneracy of the f electron state is m = 14, and n is the number of ligand atoms on which excitons are considered.

The valence band width of ligand states is neglected. In the calculations we restricted the basis set to states $|f^l x^k\rangle$ with l=0,1,2 (i.e. f^0, f^1, f^2 states) and k=0,1 (zero or one exciton/plasmon). Now the polarization energy E_p is given by

$$E_p = 1/2B(\sqrt{1 + 4nP^2/B^2} - 1) \qquad (10)$$

To illustrate the effect of the polarization energy, we calculated as an example the spectra for different values of E_p, using effectively the same parameters for ε_f, U and Q (equation 1). We took $\varepsilon_f = 0.2$, U = 7 and Q = 10.2 (all numbers are in eV). It is not possible to define an effective transfer-parameter T_{eff}[7]. so we chose T in such way that $\langle n_f \rangle^0$ was the same in all calculations ($\langle n_f \rangle = 0.62$). Results are shown in figure 1 for two distinct values of the excitation energy B. Clearly the relative intensities of the charge transfer triplet depend on E_p. In particular the intensity of the $|f^0\underline{c}\rangle$ like peak is reduced for $E_p \neq 0$. Polarization satellite peaks appear at higher binding energy.

For B = 8 we find the $|f^0\rangle$ like peak at a higher energy than the $|f^1x^1\rangle$ like satellite. Generally, however x is not a sharp state, but represents a set of states with an energy spread of several eV, as revealed for instance by electron loss spectra. So in this case (B=8) the $|f^0\underline{c}\rangle$ like final XPS

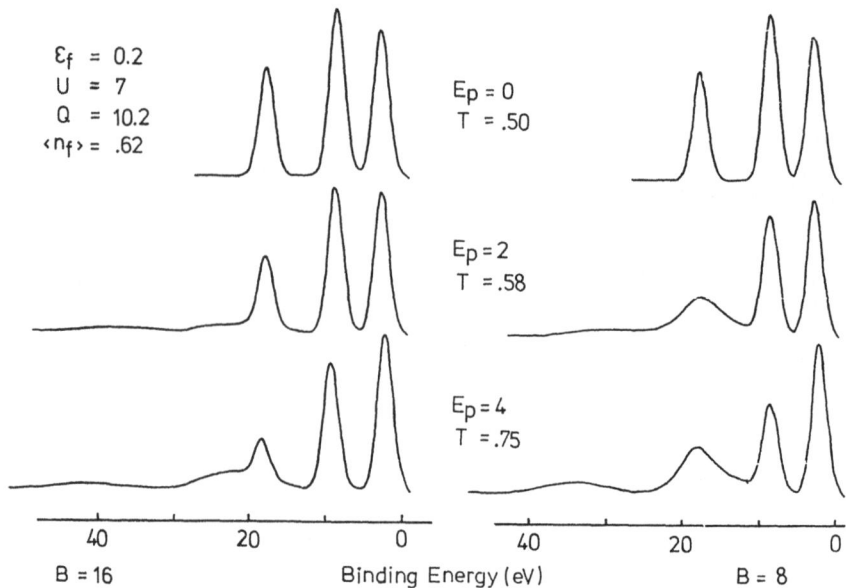

Fig. 1. Calculated core XPS spectra as a function of polarization energy E_p. (The zero of binding energy is arbitrary). Peaks belonging to the charge transfer triplet have been convoluted with a Gaussian with half width at half height w equal to 1, except for $|f^0\rangle$ like peaks for B=8 (w=2, see text). For polarization satellite peaks we used w=4.

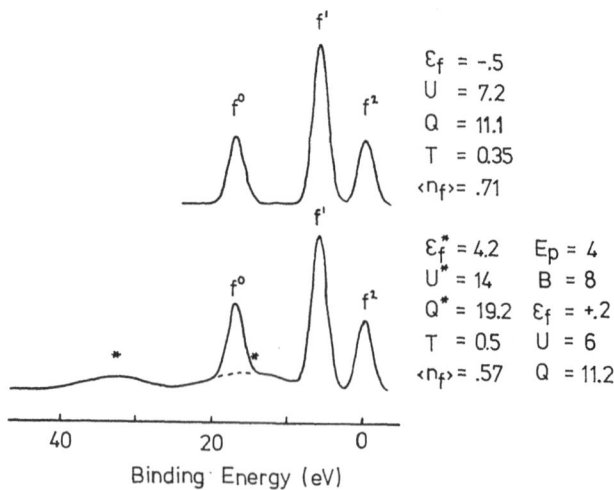

Fig. 2. Calculated core XPS spectra using parameters as indicated. Approximately the same spectrum is obtained in both cases. The main component f^1 of the peaks is indicated. * denotes satellites.

state is in fact situated in the band of $|f^1cx^1\rangle$ like states. This gives rise to a broadening of the $|f^0c\rangle$ state, in addition to the reduction of its intensity. We illustrated this in the figure by giving this peak double width. We expect that in experimental spectra often for this reason the $|f^0c\rangle$ like peak is strongly broadened or hidden by a satellite feature. In $CeAl_3$ for instance a plasmon satellite is found in the XPS spectrum in the region where the $|f^0c\rangle$ state is expected[11].

As another example we show in figure 2 the results of two calculations which lead approximately to the same spectrum. The first case is with $\varepsilon_f = -.5$ (negative ε_f:classical Ce^{III}-compound) and $E_p = 0$, resulting in $\langle n_f \rangle = .71$. The second calculation is for $\varepsilon_f = +.2$ (positive ε_f:classical Ce^{IV}-compound) and $E_p = 4$, resulting in $\langle n_f \rangle = .57$. This example shows that care must be taken in determining the value of $\langle n_f \rangle$ and the position of ε_f with respect to the Fermi level from core XPS spectra.

There are experimental indications that polarization effects are important indeed. In core XPS of several heavy-fermion Ce compounds ($CeCu_6$, $CeCu_2Si_2$, $CeH_{2.7}$) hardly any intensity of the $|f^0c\rangle$ like peak is found[12]. This could be related to a reduction of intensity due to polarization. In several Ce compounds it has not been possible so far to describe all electron spectra (core XPS, XAS, valence XPS, BIS) consistently with one set of effective parameters[13]. This is also the case in uranium compounds, where strong hybridization effects are expected[14].

REFERENCES

1. J. Zaanen, C. Westra and G.A. Sawatzky, Phys. Rev. B33:8060 (1986).
2. M. Vos, D. v.d. Marel and G.A. Sawatzky Phys. Rev. B29:3073 (1984).
3. O. Gunnarson and K. Schönhammer, Phys. Rev. B28:4315 (1983).
4. J.C. Fuggle, F.U. Hillebrecht, e.a. Phys. Rev. B27:7330 (1983).
5. W.D. Schneider, B. Delley, E. Wuilloud, J.M. Imer and Y. Baer, Phys. Rev. B32:6819 (1985).
6. W. Beall Fowler and R.J. Elliot, Phys. Rev. B34:5525 (1986).
7. D.K.G. de Boer, C. Haas and G.A. Sawatzky, Phys. Rev. B29:4401 (1984).
8. W.C. Martin, R. Zalubas and L. Hagan, Atomic Energy Levels of Rare Earth Elements; Nat. Bur. Stand. 1978.
9. S.H. Liu, Phys. Rev. Lett. 58:2706 (1987).
10. K. Schönhammer and O. Gunnarson, Phys. Rev. B30:3141 (1984).
11. H.R. Ott, Y. Baer and K. Andres in: "Valence Fluctuations in Solids" (ed. L.M. Falicov) 1981.
12. L. Schlapbach, S. Hüfner and T. Riesterer, J. Phys. C19:L63 (1986).
13. J.W. Allen, this conference.
14. D.D. Sarma, F.U. Hillebrecht, W. Speier, N. Martensson and D.D. Koelling, Phys. Rev. Lett. 57:2215 (1986).

DISCUSSIONS ON HIGH ENERGY SPECTROSCOPIES

The dual aims of this group were to establish "the conventional wisdom" on how reliable were the contributions of high energy spectroscopies (HES) to the science of narrow band materials and to discuss what still had to be done. However no one present could, or would, present arguments that HES results were not relevant to ground state properties. It was rather forcefully stated that there had been some misunderstanding in the past because people had not always understood that the size of the perturbation of the experiment was usually more relevant than the time scale. It was also stated that in both low energy and high energy scale experiments one was probing transitions between the same stationary Eigenstates of the system and that there could be no question that the results from the different energy scales were not related.

After getting rid of pent-up frustrations on that point the group went on to discuss more quietly the values for the effective correlation energies coming out of HES. It was agreed that these values were, with hindsight, generally quite acceptable to people studying near ground state properties. Also it was agreed that the remaining uncertainties in their evaluation were generally not such as to invalidate the main conclusions. On the one hand more precise interpretation of HES data was necessary for tests of consistency between different experiments, but on the other hand some people expressed the view that existence of a "unified theory" could only mean that much of the important physics, associated with the different experiments, was left out.

A fundamental input to all present day treatments of HES is the "Sudden Approximation". There has been very little experimental work to justify this approximation and there is some evidence of significant discrepancies. There has been even less theoretical work done in this area. The success of HES in the field of narrow band materials is however itself empirical evidence that the approximation is reasonable. There is no choice but to proceed on the basis of the sudden approximation, but this should present an opportunity for a bright theorist[1].

In detailed discussions the effect of lattice relaxation was considered. This is not relevant in HES, but may occur in some low energy scale experiments. We could only say that lattice relaxation should never change measured U by more than one eV and that there were many systems where one eV would not significantly affect the implications of HES results. If and when systems were investigated in which a one eV uncertainty in U was important, then one would have to consider lattice relaxation more carefully. The renormalization of U by hybridization was considered an important effect. In view of its effect, one should be cautious in statements about the influence of chemical environment on U. Also as we learn more about this we will probably have to be more precise about "what U is meant" in some discussions.

There was much discussion on what should be added to the present Anderson Hamiltonian treatments. They had generally worked well for most Ce systems, although it was difficult to fit the spectra from some isolated systems, like $CeRh_3B_2$. It is clear that the effects that should be included are

a)   The Coulomb interaction between the conduction electrons and the local electrons (i.e. d-f interactions in lanthanides). This is at present implicitly included in most Anderson Hamiltonian treatments, e.g. for core level XPS by defining a Uac which reflected the **difference** between the core hole-f and the core hole-d interactions.

b)   The Coulomb interaction between the conduction electrons (i.e. d-d for lanthanides). These are generally considered less important than the interactions between localized and conduction electrons. d-d interactions cannot always be included along with the d-f interactions.

c)   Multiplet effects. It is not terribly difficult to include final state multiplet effects by adding together contributions with dif   ent U, but there are real problems when the ground state has more than one localized electron or hole because one is then forced to go to much higher order in 1/N.

d)   Modification of the Anderson model parameters by the core hole in spectroscopies involving creation or annihilation of a core hole.

e)   Anderson lattice effects. Most treatments published up to now used the Anderson Impurity Hamiltonian and their success for Ce compounds suggested that the approximation was not too bad. However, there was no guarantee that this would also be the case for actinides or transition metal compounds.

In addition to these points questions were raised by Folkerts and Haas about polarization effects (see their contribution). The point was also raised and accepted, that much of the work on narrow band systems with high energy spectroscopies involved interpretation of satellite intensities, but we should not lose sight of the fact that satellite peaks also occur in more simple systems (see e.g. the contribution of Mårtensson).

Discussions about the problems of extraction of U_{eff} from Auger spectra when the valence band was not nearly empty led only to the conclusion that this was a real problem.

There was also some general discussion about systems other than the lanthanides. First of all, it was agreed that not only the rare earths showed mixed valence phenomena. Reference was made to the superconducting intercalation compound $SnTaS_2$ which shows XPS peaks consistent with a $Sn^0(5s^25p^2)$--$Sn^{2+}(5s^2)$ fluctuation. Then there are compounds of uranium and the actinides. Here nobody was confident that the methods applied to the lanthanides would be applicable to U but it was agreed that somebody should try. It was agreed that at least the multiplet interactions (c above) would probably have to be included. For the transition metal compounds it was agreed that methods based on the Anderson impurity Hamiltonian are probably applicable to the more ionic large band gap compounds like the late 3d transition metal halides. However, the difficulties increase as one goes to the more covalent, small band gap and metallic systems where eventually a new approach will be needed.

REFERENCES

1.   For a study of satellite intensities as a function of photon energy. See e.g. T.A. Ferrett, D.W. Lindle, P.A. Heimann, W.D. Brewer, U. Becker, H.G. Kerkhoff and D.A. Shirley, Phys. Rev. A36:3172 (1987), and references therein. For the photoenergy dependence of plasmons in solids see J.C. Fuggle, R. Lässer, O. Gunnarsson and K. Schönhammer, Phys. Rev. Lett. 44:1090 (1980).

D

HIGH TEMPERATURE
SUPERCONDUCTORS

SECTION D: HIGH TEMPERATURE SUPERCONDUCTORS

INTRODUCTION

It was appropriate to consider high temperature superconductors at this meeting because it is already clear that the Cu d electrons are strongly correlated and that many of the questions being asked about the copper perovskites are the same as those asked about NiO or CuO. To understand the relevance of the present developments in high temperature superconductivity it is desirable to recall some of the earlier developments. Although superconductivity was discovered in 1911,[1,2] the first really important insight into the phenomena came in the 1930's after the discovery of Meissner and Ochsenfeld that magnetic flux lines were actively expelled from a sample when the material in a magnetic field was cooled through the transition to the superconducting state[3]. This led to the concept that diamagnetism was a fundamental characteristic of the superconducting state, and to the London equations[4,5].

The interpretation of these equations is that the superconducting state wave function is very rigid and macroscopic in dimensions. Whilst the London equations cannot be rigorously derived, they are taken as the foundation for all theoretical descriptions of the superconducting state. Indeed much of the present theoretical work is based on the search for descriptions of the new materials which are adequate for the specific material characters (band structure, on-site Coulomb correlations, ligand metal transfer energies, etc.) but which satisfy the London criteria for the superconducting state.

Much of the work on superconductivity between 1935 and 1955 was based on thermodynamic relationships. An important step was presented in Ginsburg and Landau's ideas[6]. They defined a parameter psi relating the London penetration depth (λ_L) and the coherence length ξ. Stated briefly, λ_L is the depth to which a magnetic field can penetrate into a superconducting solid, and ξ^3 is the minimum volume in which an average value of the superconducting current, J_s, can be defined (for further details see references 6-9). Ginsburg related his parameter to the free energy of the boundary between the normal and the superconducting state. If ξ/λ_L is less than $\sqrt{2}$ the free energy is positive and the interfacial surface will be minimized. These are type I superconductors, in which the superconducting state is seen to collapse quite suddenly as the magnetic field is increased. If $\xi/\lambda_L < \sqrt{2}$ the normal-superconducting boundary has negative free energy and the area of the boundary layer is maximized when the normal and superconducting states coexist in equilibrium. As a consequence, when the magnetic field around a superconductor is increased beyond a critical value, Hc_1, where the magnetic flux lines begin to penetrate the solid, the material develops filaments of normal material, each with a single magnetic flux quantum. As the field is increased the proportion of normal material increases, until at a critical value, Hc_2, superconductivity disappears altogether. This behaviour is known as type II superconductivity (see Table I for data on ξ/λ_L).

The major step forward in the theory of superconductivity was, of course, the theory of Bardeen, Cooper and Schrieffer (BCS) in 1957.[7] The BCS theory is a model Hamiltonian set up to explain the following characteristics of the superconducting state:

1) It is a fundamental state of matter which cannot be derived from the normal solid state using perturbation theory.
2) It exists (in a minority of materials) for temperature $T<T_c$ and is reached by a second order transition.
3) It has an electronic specific heat varying as $\exp(-T_0/T)$ near $T=0$ K and other evidence for an energy gap for individual particle like excitations.
4) The Meissner-Ochsenfeld effect.
5) Effects associated with vanishing electrical resistivity.
6) The critical temperature often shows an isotope effect so that $M^\alpha Tc =$ constant, where $\alpha = 1/2$ in BCS in its original form. However, values of α in the literature range from ≈ 0 to ≈ 1 and the isotope effect is not always detectable.
7) The superconducting state wave functions satisfy the London criteria.

TABLE 1. London penetration depths, λ_L and coherence lengths ξ in Angstroms, and Tc for various superconductors (data from ref. 8-13).

Material	ξ	λ_L	ξ/λ_L	$Hc_2(T)$	Tc (K)
Sn	2300	340	6.2	0.03	3.72
Al	16000	160	100	0.01	1.14
Pb	830	370	2.2	0.08	7.19
Cd	7600	1100	6.9	0.003	0.56
Nb	380	390	1.03	0.2	9.5
Nb_3Sn				≈ 24	18.05
Nb_3Ge					23.2
$PbMo_6S_8$	23			60	≈ 14
UPt_3	200	3600	0.055		
$BaPb_{25}Bi_{.75}O_3$	60-70	5000-	≈ 0.01	≈ 0.6	11-12
$LiTi_2O_4$					≈ 12.4
$(LaSr)_2CuO_4$	23				≈ 40
	35-50				
$Yba_2Cu_3O_7$	5-40*	1000-8000*	<0.01	≈ 200	≈ 92
	22-34+	200-100+	<0.1		

* = along c crystallographic axis   + = in ab plane

At the time the BCS theory was formulated, it was already known that some sort of attractive force between the electrons was necessary to explain the energy gap. Magnetic and exchange interactions had been rather unfashionable as the attractive force since the debate between Heisenberg[14] and London[5] in the late 1940's, while there was a great deal in favour of electron-phonon interactions as the medium for the attraction (see e.g. ref. 15-17). The discussion of the original BCS papers is strongly directed towards electron-phonon interaction as the source of the attractive force of interaction, but the BCS formalism is certainly valid for other forms of interactions.

The BCS theory is based on a model Hamiltonian and cannot predict values of T_c or for real superconductors. However, Eliashberg and McMillan derived two highly non-linear expressions which can be coupled and solved by iteration to yield T_c for various scenarios[18,19]. It was generally considered that T_c could not exceed 35-40 K if electron-phonon interaction led to electron attractions. This became known as the "McMillan limit" and for nearly 20 years people thought they were safe to pontificate on this maximum. Note that we failed to find any reference to 35-40 K in McMillan's paper and we do not know precisely where it came from.

Most commercial applications involve the use of the high current carrying capability which is maintained in type II superconductors even in high magnetic fields. The most promising materials for these applications were, until recently, the A15 compounds like Nb_3Sn, Nb_3Ge and V_3Ga. In the future a special role was projected for the Chevrel-phase compound based on $PbMo_6S_8$ which have upper critical fields of up to 60 Tesla[13]. The new high T_c materials have unprecedentedly high Hc_2 values of the order of 200 Tesla (see Table I).

As shown in Table I, the best data for the new superconductors at present available shows very short coherence lengths and quite large penetration depths! They are known to be type II. Note, however, that although the coherence lengths and penetration depths of the new superconductors contrast with those of most traditional superconductors, they are not dissimilar to those found for heavy fermion superconductors such as UPt_3. This has led some people to wonder if there may be some similarities in the underlying mechanism of superconductivity of heavy Fermion and high T_c superconductivity.

Early examples of superconductivity were mostly found in materials which we think of as normal metals. It used to be considered that superconductivity and strong magnetism were mutually exclusive, and as discussed earlier in this book, strong magnetic phenomena are normally associated with narrow bands. One would thus not initially have expected superconductivity in oxides, where narrow band effects are common. The first oxide in which superconductivity was found was $SrTiO_3$ doped with Nb, in which T_c can be up to 0.5 K [10]. $SrTiO_3$ itself is an almost ferroelectric insulator with an extremely high dielectric constant. Contrary to the high T_c materials, the charge carriers in the Nb doped material are probably electrons[10]. Superconductivity in $LiTi_2O_4$ ($Tc \approx 12$ K) was first observed by Johnston et al.[11]. This material has a spinel crystal structure with "mixed valent" Ti (Ti^{III} and Ti^{IV}) and probably conduction via electrons in the Ti 3d band. $Ba(Pb_{1-x}Bi_x)O_3$ is another example of a superconductor for x < 0.3 with Tc up to 12 K [12], in which the conduction is via holes although the character of the holes is not known. For x > 0.3 these oxides are semiconductors with small activation energies for conductivity. Although we do not know if there is a connection with the high T_c materials as far as the mechanism for superconductivity is concerned, these oxides are similar to the new superconductors in that all (except perhaps $Ba(Pb_{1-x}Bi_x)O_3$) are narrow band systems.

The new family of superconductors clearly do exceed the old McMillan limit. In addition they have the following properties:

1. All are pseudo ternary or quarternary compounds of:

    Y   Ba   Cu   O
    La   Sr
    R.E.   Ca

    They often lack simple stoichiometry.

2. They are all anisotropic-layered solids in which the superconductivity seems to be within the "CuO_2" layers.

3. Resistivity = 0 below very high T_c values; e.g. $La_{1.85} Sr_{0.15}CuO_4$ $\approx 36°$, $Yba_2Cu_3O_7$ ≈ 92 K.

4. Measurable Meissner effects.

5. Type II superconductivity with small coherence lengths, large Hc_2.

6. ^{16}O - ^{18}O isotope effects seen for the $LaCuO_4$ family, but not the $Yba_2Cu_3O_7$ family.

7. Traditional flux quantum ($\pi hc/2e$), as for "normal" Cooper pairs.

8. $LaCuO_{4-y}$ shows antiferromagnetic ordering for very small values of y[20].

9. An energy gap in the single particle excitation spectrum in the super-conducting state. There is some argument about its size but it may well be ≈ 3.5 K_bT_c as in the traditional BCS superconductors.

10. La_2CuO_4 and $Yba_2Cu_3O_{6.5}$ both seem to show a large gap in the density of states in the normal solid states. This gap is ≈ 2 eV in La_2CuO_4[21] so that the basis material is an intrinsic semiconductor, in contrast to the predictions of most band structure calculations[22].

Now the characteristic of the new superconductors that has attracted most attention is the high value of T_c. Most researchers in the field feel that this requires an explanation. In our discussions Prof. Leo Falicov put forward a classification of the explanations which is the basis of the following.

Class I
An ordinary superconducting BCS state which, for some specific reason exceeds the "McMillan limit", e.g.;
a) very high Debye cut off
b) very peculiar α^2F, where $\alpha(\omega)$ is the phonon coupling constant and $F(\omega)$ is the phonon spectrum.
c) strange density of states
d) low dimensionality (anisotropy)
e) specific soft modes
f) near instability
g) etc.

Class II
A superconducting BCS state in which the attractive force is not phononic in origin, e.g.;
a) plasmons
b) excitons
c) dielectric polarization

d)   d-wave plasmons
e)   magnons
f)   negative U centres where U is the effective on-site correlation energy
      (see contribution from C. Haas, this book).
g)   etc.

Class III
A superconducting BCS state in which the fermions are "strange" quasiparticles (e.g. heavily dressed electrons of a non-ordinary metal) e.g.:
a)   heavy Fermions
b)   resonant-valence-bond semi-metal
c)   mixed-valent antiferromagnetic (double exchange)
d)   etc.

Class IV
A superconducting ground state altogether different from that proposed by
BCS, e.g.:
a)   Bose condensation of polyfermion bosons (e.g. bi-excitons, bi-polarons)
b)   etc.

Clearly, it is rather difficult to decide between all these possibilities. Note that most are consistent with the BCS scheme. In the course of
the summer of 1987 a concensus is accumulating that there is little strong
evidence incompatible with BCS (although this may still change). However,
a lot of theoretical work is being stimulated. There are a lot of questions
which can be asked which are related to those normally asked by scientists
in the field of narrow band phenomena. e.g. How large are the values of
U_{eff}, the effective Coulomb correlation energies between the Cu d and O 2p
levels? Are they so large that one electron band theory becomes of limited
use? How large are the band widths and the transfer integrals and how strong
are hybridization effects? Do the very short coherence lengths imply that
the attraction between the charge carriers is short ranged? What is the
character of the holes that lead to superconductivity? Is the Cu^{3+} ($3d^8$)
configuration important or are the holes in the O2p bands? If so, which
O2p bands?

REFERENCES

1.   H. Kamerlingh Onnes, Leiden Comm. 124C, (1911).
2.   H. Kamerlingh Onnes, Akad. van Wetenschappen (Amsterdam) 14:113, 818
      (1911).
3.   W. Meissner and R. Ochsenfeld, Naturwiss 21:787 (1933).
4.   F. London and H. London, Proc. Roy. Soc. A149:71 (1935); A152:24 (1935).
5.   F. London, Phys. Rev. 74:562 (1948).
6.   V.L. Ginsburg, JETP (USSR) 14:134 (1946); Uspekhi Fiz Nauk 48:25 (1952).
7.   J. Bardeen, L.N. Cooper and J.R. Schrieffer, Phys. Rev. 106:162 (1957);
      108:1175 (1957).
8.   C. Kittel, in: "Introduction to Solid State Physics", 5th Ed. J. Wiley,
      New York (1976).
9.   "Superconductivity in Ternary Compounds", Ed. O. Fischer and M.B.
      Maple, Springer, N.Y. (1982); "Proceedings of the Adriatico Conference
      on High Temperature Superconductors", Eds. S. Lundqvist, E. Tosatti,
      M. Tosi and Yu Lu, World Scientific, Singapore, (1987); J. Franse,
      private communication.
10.   H.P. Frederikse, J.F. Schooley, W.R. Thurber, E. Pfeiffer and W.R.
      Holser, Phys. Rev. Lett. 16:579 (1966).
11.   D.C. Johnston, J. Low Temp. Phys. 25:145 (1976); R.W. McCallum, D.C.
      Johnston, C.A. Luengo and M.B. Maple, J. Low Temp Phys. 25:177 (1976).
12.   A.W. Sleight, J.L. Gillson and P.E. Bierstedt, Solid State Commun.
      17:27 (1975); B. Batlogg, Physica 126B;275 (1984) and references
      therein.

13. O. Fischer, H. Jones, C. Bongi, M. Sergent and R. Chevrel, $\underline{J}$. $\underline{Phys}$. $\underline{C}$ 7:L450 (1974).
14. W. Heisenberg, $\underline{Z}$. $\underline{Naturforschung}$ 2a:424 (1947).
15. N. Bohr, $\underline{Physica}$ 19:761 (1953); H.B.G. Casimir ibid 19:764 (1953).
16. H. Fröhlich, $\underline{Phil}$ $\underline{Mag}$. 41:221 (1950); $\underline{Phys}$. $\underline{Rev}$. 79:845 (1950).
17. J. Bardeen, $\underline{Rev}$. $\underline{Mod}$. $\underline{Phys}$. 23:261 (1951).
18. G.M. Eliashberg, $\underline{Sov}$. $\underline{Phys}$. $\underline{JETP}$, 11:696 (1960).
19. McMillan, $\underline{Phys}$. $\underline{Rev}$. 167:331 (1968).
20. S. Mitsuda, G. Shirane, S.K. Sinha, D.C. Johnston, M.S. Alvarez, D. Vaknin and D.E. Moncton, $\underline{Phys}$. $\underline{Rev}$. B 36:822 (1987).
21. see e.g. J. Orenstein, G.A. Thomas, D.H. Rapkine, C.G. Bethea, B.F. Levine, B. Batlogg, R.J. Cava, D.W. Johnson Jr. and E.A. Rietman, $\underline{Phys}$. $\underline{Rev}$. B36:8829 (1987); K. Kamaras, C.D. Porter, M.G. Doss, S.L. Herr, D.B. Tanner, D.A. Born, J.E. Greedman, A.H. O'Reilly, C.V. Stager and T. Timusk, $\underline{Phys}$. $\underline{Rev}$. $\underline{Lett}$. 59:919 (1987); S. Tajima et al. $\underline{Jpn}$. $\underline{J}$. $\underline{Appl}$. $\underline{Phys}$. 26:L432 (1987).
22. L.F. Mattheis, $\underline{Phys}$. $\underline{Rev}$. $\underline{Lett}$. 58:1028 (1987); J. Yu, A.J. Freeman and J.H. Xu, $\underline{Phys}$. $\underline{Rev}$. $\underline{Lett}$. 58:1035 (1987); K. Takegahara, H. Harima and A. Yanase, $\underline{Jap}$. $\underline{J}$. $\underline{Appl}$. $\underline{Phys}$. 26:L352 (1987); R.A. de Groot, H. Gutfreund and M. Weger, $\underline{Solid}$ $\underline{State}$ $\underline{Commun}$. 63:451 (1987); W. Temmerman, G.M. Stocks, P.J. Durham and P.A. Sterne, $\underline{J}$. $\underline{Phys}$. $\underline{F}$. $\underline{Met}$. $\underline{Phys}$. 17:L135 (1987).

SUPERCONDUCTING OXIDES

John B. Goodenough

Center for Materials Science & Engineering
ETC 5.160, University of Texas at Austin
Austin, TX 78712

PHASE DIAGRAMS

In a solid, the character of the electrons outside of closed atomic shells depends upon the relative strengths of four competing energies:

1. Heat: $T\Delta S$
2. Intraatomic energies: U, Δ_{LS}
3. Interatomic energies: b_{ij}
4. Electron-lattice interaction energy: Δ_{el}

Heat is responsible for solid-solid phase transitions at a transition temperature T_t or T_c as well as the solid-liquid transition at the melting point T_{mp}.

The principal intraatomic energy of interest is the energy U required to add an electron to an atom having a partially filled set of orbitals. In the free-atom or free-ion limit, it is given by the difference in successive ionization energies; for ion complexes in solution, it is given by successive redox energies. Covalent mixing with ligand orbitals in a molecule or a solid can reduce U dramatically from its free-atom or free-ion value.

The multiplet splitting Δ_{LS} can be important where the orbital angular momentum is not quenched by the crystalline fields.

Interatomic interactions give rise to a resonance (or transfer-energy) integral t_{ij}, which contains a spin-independent factor

$$b_{ij} \equiv (\Psi_i \cdot H'\Psi_j) \tag{1}$$

where H' is the perturbation of the electronic potential at R_j by an atom at R_i. In oxides, two types of interatomic interactions must be distinguished: the cation-anion nearest-neighbor resonance integral $b_{ij} \equiv b^{ca}$ and the nearest-neighbor like atom interaction energies $b_{ij} \equiv b$. In the tight-binding limit, the interaction between equivalent orbitals or like atoms gives rise to a bandwidth

$$w_b \approx 2zb \tag{2}$$

where z is the number of nearest like neighbors.

In the limit $w_b \ll U$, the electrons are strongly correlated, and the atoms have localized magnetic moments so long as a set of degenerate atomic orbitals is partially occupied (and the spin moment is not cancelled by an orbital moment). In this case, the interatomic interactions for non-orthogonal orbitals give rise to an interatomic exchange coupling that may be described by a perturbation of the localized crystal-field configuration. For a formally mixed-valent system, the double-exchange interactions are ferromagnetic and described by first-order theory; for a formally single-valent system, the superexchange interactions are antiferromagnetic where described by second-order theory, ferromagnetic where described by third-order theory.

In the limit $w_b \gg U$, the one-electron band theory is applicable, and there is no spontaneous magnetism. So long as the translational symmetry does not split a partially filled band at the Fermi energy, metallic conductivity is observed.

Where the interatomic and intraatomic energies are of comparable magnitude ($w_b \approx U$), cooperative electron-phonon interactions determine whether the electrons in partially filled bands become spontaneously magnetic, induce charge-density waves, or condense out as Cooper pairs in a superconducting state.

Electron-phonon interactions are also important in the limits $w_b \ll U$ and $w_b \gg U$. In the case $w_b \ll U$, these interactions are responsible for cooperative Jahn-Teller distortions, exchange striction, and magnetostriction; they induce small-polaron formation in mixed-valent conductors. In the opposite limit, $w_b \gg U$, electron phonon interactions induce large-polaron or Cooper-pair formation, and they scatter nearly free electrons to place a lower limit on the resistivity in the normal metallic state.

From these general considerations, it is possible to construct phenomenological phase diagrams. Fig. 1 illustrates modified Hubbard diagrams for the case of a half-filled band, and Fig. 2 represents the corresponding phase diagrams where the value of U at small bandwidth is (a) large and (b) small relative to any splitting Δ_q associated with a charge-density wave (CDW).

For a non-degenerate band, a correlation energy $U > w_b$ splits the band in two, and materials with a half-filled band are antiferromagnetic insulators at T = 0K. Where the small-bandwidth U is small, a CDW can compete with antiferromagnetic ordering by changing the translational symmetry so as to stabilize occupied states at the expense of unoccupied states. If the band is more (or less) than half-filled, a formal mixed valence makes conduction possible in the absence of a CDW; however, small-polaron formation introduces a motional enthalpy into the mobility.

For degenerate bands, a correlation splitting U is associated with each single-valent electron configuration, and in the limit $U > w_b$ the sign of the superexchange interactions and the possibility of Jahn-Teller distortions depend upon the occupancy of the degenerate bands.

APPLICATION TO OXIDES

Application of these general concepts to a specific transition-metal oxide begins with the construction of crystal-field orbitals. In cubic symmetry, the fivefold-degenerate d-electron manifold is split into a threefold-degenerate set of t_{2g} orbitals and a twofold-degenerate set of e_g orbitals. If the corresponding atomic orbitals are f_t and f_e, then in an

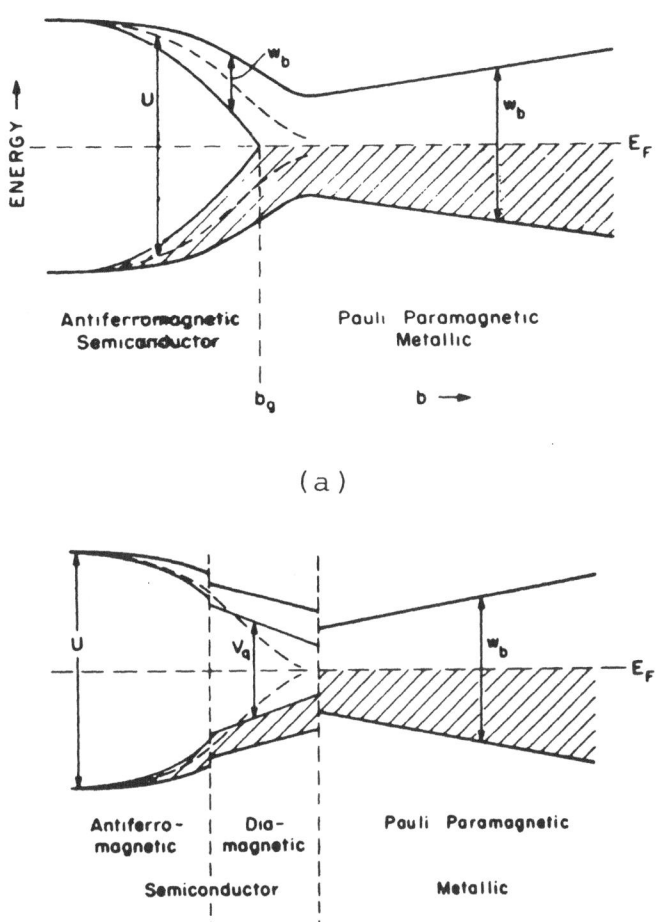

(a)

LATTICE INSTABILITIES

(T = OK, n_{ℓ} = 1, V_q > U for b'_t < b < b_t, and s.c. array)

(b)

Fig. 1. Modified Hubbard diagram for a half-filled band: (a) large U and (b) small U at small bandwidth.

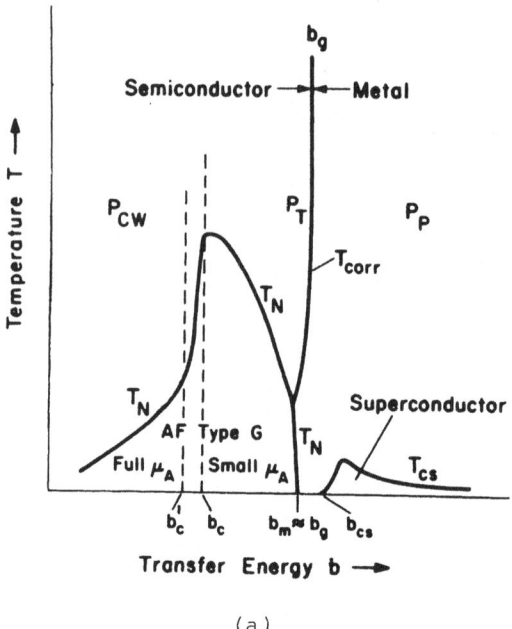

(a)

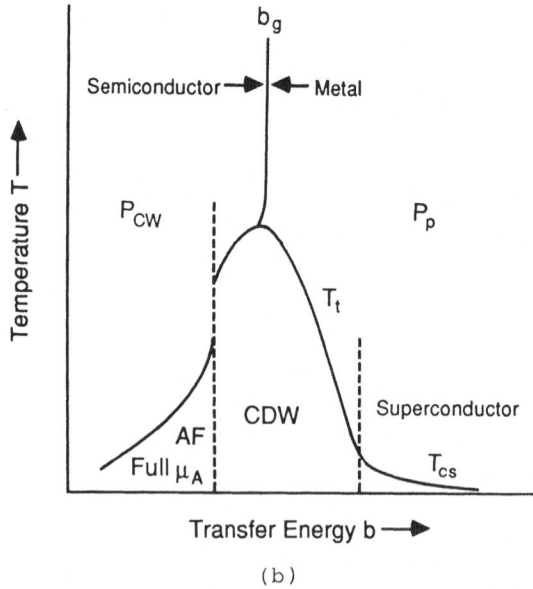

(b)

Fig. 2   Conceptual phase diagram of temperature vs. nearest
         like neighbor transfer energy b for a half-filled
         band: (a) large U and (b) small U at small b.

octahedral interstice of an oxide lattice the crystal-field orbitals become

$$\Psi_t = N_t(f_t - \lambda_\pi\phi_\pi); \quad \Psi_e = N_e(f_e - \lambda_s\phi_s - \lambda_\sigma\phi_\sigma) \tag{3}$$

where N_t, N_e are normalization constants and ϕ_π, ϕ_s and ϕ_σ are T_{2g}-symmetrized or E_g-symmetrized oxide-ion p_π, s, and p_σ orbitals (π and σ bonding). The magnitudes of the covalent-mixing parameters

$$\lambda_\pi = b_{t-\pi}^{ca}/(E_t-E_\pi), \quad \lambda_s = b_{e-s}^{ca}/(E_e-E_s), \quad \lambda_\sigma = b_{e-\sigma}^{ca}/(E_e-E_\sigma) \tag{4}$$

and the ratio of a cation-anion transfer-energy integral

$$b_{ij}^{ca} = (f_i, H'\phi_j) \approx \varepsilon_{ij}(f_i, \phi_j) \tag{5}$$

and the energy difference (E_i-E_j) of the cation acceptor orbital f_i and the anion donor orbital ϕ_j. In general, the overlap integrals have the relationships

$$(f_e, \phi_\sigma) > (f_e, \phi_s) > (f_t, \phi_\pi) \tag{6}$$

so that covalent mixing raises the energies of the antibonding e_g orbitals above those of the antibonding t_{2g} orbitals. The total cubic-field splitting is

$$\Delta_c = 10Dq = \Delta_m + (1/2)[\lambda_s^2(E_e-E_s) + \lambda_\sigma^2(E_e-E_\sigma) - \lambda_\pi^2(E_t-E_\pi)] \tag{7}$$

Once the crystal-field orbitals have been constructed, the interatomic interactions between crystal-field orbitals on like atoms can be considered. Note that covalent mixing spreads the crystal-field wave functions out over the anions, which not only reduces U but also allows metal-anion-metal interactions to become comparable to - or greater than - the metal-metal interactions. Note also that the association of formal cation and anion valences is predicted on a Fermi energy located above the top of the bonding bands of primarily anion-p character, the $0:2p^6$ bands. This condition may not be fulfilled where the cations have an unusually high formal valence state.

SUPERCONDUCTIVITY

According to the phase diagrams of Fig. 2, the superconducting state is a property of itinerant electrons with the superconducting transition temperature varying as

$$T_c \approx \Theta_D \exp(-1/V_{BCS}N(0)) \tag{8}$$

where $N(0)$ is the density of states at the Fermi energy at $T = 0K$, Θ_D is the Debye temperature, and V_{BCS} is a Cooper-pair potential that decreases with increasing correlation energy U. Therefore T_c is suppressed by U as w_b is decreased toward $w_b \approx U$. Where the correlation energy U at small bandwidth is smaller than the band splitting induced by a charge-density wave (CDW), there the superconducting state may be suppressed by the formation of a static CDW, Fig. 2(b). These considerations indicate that a high T_c is to be found only where the U at small bandwidth is smaller than the electron-phonon coupling energy for a Cooper pair and the formation of a static CDW is suppressed.

Suppression of a static CDW is more likely where the formal valence state is mixed; but the possibility of a longer-wavelength - even incommensurate - CDW remains. Until the recent discoveries of superconducting transition temperatures $T_c \geq 40K$ in $La_{2-x}Sr_xCuO_{4-y}$ and $T_c > 90K$ in $Ba_2YCu_3O_{8-y}$, it was commonly assumed that electron-phonon interactions of sufficient magnitude for high-temperature superconductivity would favour the formation of a static CDW that suppressed the superconducting state.

In this connection, it is useful to recall the compound Ti_4O_7, which is a shear structure containing reduced TiO_2 slabs between shear planes. The TiO_2 structure consists of chains of edge-shared octahedral running parallel to the tetragonal c-axis; the chains are joined by common octahedral-site corners. At low temperatures, the d electrons of the $Ti^{4+/3+}$ couple are condensed into an ordered array of homopolar bonds; at high temperatures they occupy a narrow d band. The ordered array of Ti-Ti homopolar bands represents a static CDW. In a narrow intermediate range of temperature, the homopolar bonds become disordered and mobile. These mobile electron pairs have been called bipolarons[1]. They are distinguishable from a Cooper pair because they move diffusively like a small polaron; the momentum vector $\mathbf{k}$ is not a good quantum number. Like a large polaron, the momentum vector of a Cooper pair has a defined momentum vector $\mathbf{k}$. By analogy, it is therefore possible to envisage a Cooper pair as a large bipolaron. In the formally mixed-valent spinel $Li[Ti_2]O_4$, which is superconducting below $T_c \approx 11K$ [2], the Cooper pairs may be envisaged as large, mobile bipolarons centered alternately at Ti-atom dimers and tetramers.

Mobile bipolarons associated with cation (or anion) clustering are to be distinguished from those associated with a disproportionation of the type

$$M^{m+} + M^{m+} \rightleftharpoons M^{(m+\delta)+} + M^{(m-\delta)+}, \quad 0 < \delta \leq 1 \qquad (9)$$

Only the latter can occur where the anion p orbitals are filled and only M-X-M interactions are present. For example, the cubic perovskite $CaFeO_3$ contains high-spin $Fe^{4+}:t_2^3\sigma^{*1}$ configurations, the notation σ^* signifying that the σ-bonding Fe:e - O:p_σ - Fe:e interactions are strong enough to broaden the Fe:e orbitals into a narrow σ^* band of one-electron states having a crystal-field e-orbital parentage. Takano et al.[3] observed a splitting of the Mössbauer spectrum below 290K, the two isomer shifts corresponding to a disproportionation

$$Fe^{4+}:t_2^3\sigma^{*1} + Fe^{4+}:t_2^3\sigma^{*1} \rightleftharpoons Fe^{(4+\delta)+}:t_2^3\sigma^{*(1-\delta)} + Fe^{(4-\delta)+}:t_2^3\sigma^{*(1+\delta)}$$

with δ increasing continuously from zero to essentially one with decreasing temperature. Such a charge transfer, done at the expense of U, is to be contrasted with the cooperative Jahn-Teller distortion found in $LaMnO_3$, which has a $Mn^{3+}:t_2^3e^1$ configuration. Creation of polar states by disproportionation is done at the expense of U; it is only possible where the crystal-field U is small (in this case due to σ-bond covalent mixing) and the gain in bonding energy at the oxidized species is strongly enhanced by removal of a σ-bonding electron. In the perovskite structure, a cooperative displacement of all oxygen atoms away from the reduced ion toward the oxidized ion is possible.

In $CaFeO_3$, the paired electrons occupy degenerate orbitals and are not spin-paired on a high-spin Fe^{3+} ion; also disproportionation introduces a static CDW. On the other hand, in the cubic perovskite $BaBiO_3$, disproportionation of the $Bi:6s^1$ band leaves the electrons spin-paired. However, the ordering of the bismuth represents a static CDW, so here also there are no bipolarons. On the other hand, superconductivity has been found in the mixed system $BaPb_{1-x}Bi_xO_3$ in the compositional range where a static

CDW is not formed[4]. Such observations have prompted theories of superconductivity based on disproportionation bipolarons[5,6].

A conceptual problem for superconductivity based on disproportionation bipolarons is, in my view, the mode of conduction as a bipolaronic entity. As the anions shift to trap the bipolaron at a new site, they must move through the mid position of the cation-cation axis where the charge should be evenly distributed; therefore, the electrons should jump sequentially between sites rather than as a bipolaronic unit. This problem may be alleviated if the anion-p bands overlap the Fermi energy, and indeed $BaPbO_3$ is a semi-metal.

These considerations will be discussed with reference to the new superconducting oxides $La_{1-x}Sr_xCuO_{4-y}$ and $Ba_2YCu_3O_{8-y}$.

REFERENCES

1. B.K. Chakraverty, *Jour*. *de* *Phys*. *Lett*. 40:L-99 (1979).
2. D.C. Johnston, *J*. *Low-Temp*. *Phys*. 25:145 (1976).
3. M. Takano, N. Nakanishi, Y. Takeda, S. Naka, and T. Takada, *Mat*. *Res*. *Bull*. 12:923 (1977).
4. A.W. Sleight, J.L. Gillson, and F.E. Bielstedt, *Solid* *State* *Commun*. 17:27 (1975).
5. P. Prelovsek, T.M. Rice, and F.C. Zhang, personal communication.
6. W. Weber, *Phys*. *Rev*. *Lett*. 58:1371 (1987).

WHAT CHEMISTRY IS THERE IN HIGH TEMPERATURE SUPERCONDUCTIVITY?

LOW ENERGY SCALE PROPERTIES OF NARROW BAND SYSTEMS

John A. Wilson

H.H. Wills Physics Laboratory
University of Bristol
Bristol BS8 1TL, U.K.

By chemistry above is meant a detailed appreciation of the structural, energetic and bonding characteristics of the individual superconducting material in relation to the full and appropriately graded array of other real materials[1]. This appreciation must go way beyond the simple e/a counts (along with cubic structure) that were for so long advocated by Matthias as being what counts in superconductivity[2]. That was largely a consequence of what materials physicists were prepared to take to low temperatures, or rather, what materials it proved possible for physicists and metallurgists to prepare without employing 'chemical techniques'. Those days are now overtaken by the enthusiastic study of materials like black phosphorus, or $PbMo_6Se_8$, or $(BEDT-TTF)_2I_3$, or $YBa_2Cu_3O_7$.

Superconductivity runs through the whole gamut of metals; among the elements from Hg and Nb to U and to high pressure silicon and black phosphorus[3]; or among alloys from Tl-Pb-Bi and NbTi through WBe_{13} to Nb_3Ge, at which point the term compound becomes more appropriate. With compounds the subject proceeds through pnictide superconductors such as Nb_5Sb_4 or RuSb (or VN); through chalcogenide superconductors like $TaSe_3$ or NbS_2 (or NbO); and even possibly to some halide metal like ScCl or LaI_2, to balance the score with the organic superconductor mentioned above. The new high temperature superconducting oxides indeed bring us right to the ionic limit of the metallic regime. Here the copper compounds of interest butt up against Curie-Weiss paramagnetic Mott insulators like d^8 La_2NiO_4[4], or indeed $LaSrCuO_{4-x}$[5] (held at $Cu^{2+}(d^9)$ through its oxygen defect content). It is similarly found that LiV_2O_4 with its mixture of V^{4+} and V^{3+} is a magnetic Mott insulator, and hence without the special interest of $LiTi_2O_4$, - its superconductivity at 13 K[6].

Over this full array of metals it has long been recognized that the better superconducting metals, and in particular the subclass of strong coupling superconductors, are in fact not the best of metals but rather poor metals. See for example the schematic phase diagrams presented by Goodenough in the 1960's[7], drawn up with the gradations amongst the ternary perovskites and spinels particularly in mind.

Several binary 4d and 5d oxides such as ReO_3 and RuO_2 are excellent metals because of the relatively small lattice parameter afforded by the O^{2-} ion. These are non-superconductors. However, a very large number of compound metals such as $NbSe_2$ or $RhSe_2$ or Mo_6Se_8 or La_3S_4 have resistivities which have risen to ~100 $\mu\Omega$-cm or more by room temperature and then tend to saturate at raised temperatures[8]. Many of these are good superconductors. In metals like the latter the mean free path is becoming comparable to the cell parameter, and the relaxation times rise towards the phonon periods. This tendency for $d\rho/dT$ to fall towards zero, or even turn negative at high temperatures, does not of itself betoken superconductivity as more likely. Actual disordering of Nb_3Ge by irradiation while raising ρ drives T_c rapidly downward[9]. Indeed any structural disorder in general is deleterious for superconductivity in intrinsically poor metals, although it may lift a good metal like Al into the raised T_c regime[10]. In poor metals disorder inevitably leads to loss of coherence and towards the production of local moments. Spin fluctuations or paramagnons show no evidence, as far as I have seen, for being beneficial to superconductivity: $4d^7RhSe_2$ is a superconductor, $3d^7CoSe_2$ is not[11]. Where magnetic moments and superconductivity coexist in a compound they are invariably clearly centered on different sublattices: e.g. $ErMo_6Se_8$ or $ErRh_4B_4$[12].

Magnetic moments act as pair breakers in superconductors (cf. Fe_xNbSe_2 vs. Al_xNbSe_2)[13]. Other narrow band phenomena also can run counter to the development of a superconducting groundstate, but in a somewhat different manner. Any phenomenon whose action is to gap the Fermi surface of the metal or redefine the crystal geometry so as to reduce the density of states close to E_F (and so the overall energy) is a phenomenon in competition with superconductivity.

A very robust locally-based phenomenon like metal-metal bonding thus runs strongly counter to the appearance of a superconducting ground state, for example in $RhAs_2$ or NbO_2 or NbS_3, or WTe_2 or γ-ZrI_2. Narrow band compounds may also pre-empt a metallic ground state through cluster formation (e.g. $ReSe_2$, α-ZrI_2 ($\equiv Zr_6I_{12}$)) or through adoption of some coordination that leads to a band gap at E_F (e.g. square-planar in PtS[14], trigonal prismatic in MoS_2[15], or octahedral in PtS_2[16]). Disproportionation also can gain the same end, as with $PtTeI$ ($Pt^{4+} + Pt^{2+}$)[17]. The latter phenomenon is common among non-transition metal compounds, e.g. tetragonal $InTe$[18] ($In^{3+}(s^0) + In^{1+}(s^2)$) or $BaBiO_3$[19] ($Bi^{5+}(s^0) + Bi^{3+}(s^2)$).

Disproportionation and metal-metal bonding each can be viewed as extreme forms of a commensurate charge density wave state. In the former case the charge extrema fall at the atom sites, while in the latter the extrema fall between the atoms in bonding and anti-bonding fashion[20]. I have tended to preserve the term 'charge density wave' for cases of charge redistribution and the accompanying periodic lattice distortion in which the magnitude of the changes is not so large as completely to reshape the band structure at energies well away from the Fermi level. Comparison of the very different band structures of VO_2 in its simple and its dimerized forms[21] should be contrasted with the band-folding type changes calculated for the $3a_0$ CCDW in 2H-$NbSe_2$[22].

Even for the latter material it is very hard to assess the average gapping to be associated with the CDW from the superlattice band structure. It is certainly overstated by taking the energy of maximum change in the infra-red reflectivity. As a rough guide $2\Delta_{CDW}$ is given by a BCS-like value of 3.5 kT_o. Fermi surface gapping under 'nesting' is not necessarily complete over the entire or even the major part of the Fermi surface in a CDW material like 2H-TaS_2, though with the further reduced dimensionality of o-TaS_3 the gapping does become total, and a semiconductor results at low temperatures.

For $NbSe_2$ below 5 K the gapping of the CDW (T_o 33 K) becomes subsumed in the gapping over the entire Fermi surface of its superconducting state (T_s = 7 K). This is not unlike the situation in Nb_3Sn or V_3Si where a slight distortion in the A-atom chains with symmetry breaking may or may not preceed superconductivity, dependent upon sample quality[24]. In these A15 materials the lattice distortion is again more sensitive to disorder than the superconductivy. Its presentation as a band Jahn-Teller effect would implicate some specific features in the band structure that produces a peak in the density of states very close to E_F. This is removed by the modification in symmetry and interactions which the distortion entails.

The term 'band Jahn-Teller distortion' has clearly been frequently misplaced. Take for example the ℓ.t. state in $TiSe_2$, where the form of the distortion developed emphasizes the electron-hole coupling in this semi-metal[25]. Conversely certain materials like $d^{1/2}(Ta(Se_2)_2)_2)$ are spoken of as being CDW materials, whilst in fact the adopted distortion is a transverse one that achieves Fermi surface gapping through the drop in symmetry[26]. A similar transverse distortion also seems to be responsible for the non-metallic nature of $BaVS_3$[27] and β-$TiCl_3$[28] at low temperatures. In neither of these d^1 chain materials is there any evidence of dimerization.

Clearly all systems at this level of 'energy adjustment' are highly involved. One recalls that invariably the non-metallic character of ℓ.t. VO_2 is attributed to dimerization, yet in lightly doped VO_2:Cr where half the chains are unpaired, and where there is lateral atomic shift that actually increases the V-V distance, the result is still a non-metal but of reshaped unit cell and symmetry[29].

Dimerization actually can still show up within the Mott insulating regime (e.g. with β-$RuBr_3$, or certain cupric salts[30]), and is automatically then unrelatable to Fermi surface nesting, as has been attempted for VO_2[31].

When one surveys crystal structures it is clear that a distortion type often persists over several electron counts, and hence cannot be Fermi surface specific, e.g. the MnP-type distortion of the NiAs structure in CrAs, MnAs, FeAs and CoAs[32]. Much more justification is then required before calling the similar distortion in VS a CDW, - or the recently discovered superlattice shown by NiAs itself[33].

Certain atomistically centred effects are by contrast more straightforward to identify. The distorted semiconductive nature of CrS and CrSe is due to on-site Jahn-Teller distortion[34] (here of the NiAs structure), which is endemic to all high-spin $d^4Cr^{2+}(t_{2g}^3e_g^1)$ compounds, such as CrI_2 or Rb_2CrCl_4[35], or to cupric compounds like $CuCl_2$, K_2CuF_4 or La_2CuO_4[36]. Corresponding to this is the stability often to be gained at d^2 from trigonal-prismatic coordination (e.g. in semiconductive 2H-MoS_2, etc.[15]) or at d^8 from square-planar coordination, encountered both in broad band (PtS) and narrow band compounds ($PdCl_2$)[37]. The latter configuration is so favourable that it assures semiconductivity to "d^7" PtI_3[38] and "d^9" $CsAuCl_3$[39], through structurally accommodated disproportionation (as brought semiconductivity to InTe and $BaBiO_3$).

Configurational conversions can greatly moderate the lattice condition. In octahedral coordination t_{2g} orbitals are non-bonding, e_g orbitals anti-bonding. Because of this the high-spin condition $t_{2g}^3e_g^2$ confers on $3d^5$ $FeCl_3$ a larger molecular volume than does low-spin t_{2g}^5 on $4d^5$ α-$RuCl_3$ in the same structure type. One may further recall the enormous volume contraction between f^6 black SmS and f^5d^1 golden SmS, where both states are rocksalt in structure[40].

The more complex the unit cell the more difficult it is to venture to say what is afoot. Why does $NbTe_4$ have a $3a_0$ distortion?[20b] How is the 1q state of $NbTe_2$ to be described, - or its $\sqrt{19}\ a_0$ 3q CDW? Are the changes wrought by the $\sqrt{13}\ a_0$ state of 1T-TaS_2 so severe that one should forego the CDW FS-driven description of events, and turn to a cluster description? The modification in instability wavevector through the mixed system 1T-$(Ta/Ti)S_2$ would say the former[41]. Clearly, however, in all these narrow-band phenomena the electronic and phononic/lattice aspects to the problem are inextricably mixed[42]. Can this be any less so with superconductivity? While superconductivity was a low temperature phenomenon, with a BCS gap of only 20 cm^{-1} or less, the coupling situation seemed clear-cut, although strong couplers like VN or Nb_3Ge clearly began to create problems for the traditional phononic interpretation of the Cooper-pair formation[43]. In the new superconductors like $YBa_2Cu_3O_7$, where T_c has become a significant fraction of θ_D, we are in a range of energies where the onset of superconductivity could well promote a structural/electronic change, as magnetism can in say α-NiS[44] and V_2O_3[45].

Can we now bring some of the above points to bear on how to interpret the general behaviour of the new superconducting materials, and even their superconductivity?

REFERENCES

1. a) J.A. Wilson, Adv. in Phys. 21:143 (1972).
   b) J.A. Wilson, Proceedings of the Conference on 'Phase Transitions' Ed.: L.E. Cross, Pergamon Press, p. 101-117 (1972).
   c) J.A. Wilson, NATO ASI, B 113:708 (1984). 'Electronic Structure of Complex Systems' Ed.: P. Phariseau and W.M. Temmerman, Plenum, N.Y.
   d) J.A. Wilson, Chapter 9, page 215-260 in 'The Metallic and Non-Metallic States of Matter' Ed.: P.P. Edwards and C.N.R. Rao, Taylor and Francis (1985).

2. J.K. Hulm and B.T. Matthias, NATO ASI Ser. B 68: chapter 1 (1981), ed.: S. Foner and B.B. Schwartz, Plenum.

3. Y. Takao, H. Asahina and A. Morita, J. Phys. Soc. Jap. 50:3362 (1981).

4. M. Sayer and P. Odier, J. Sol. St. Chem. 67:26 (1987).

5. N. Nguyen, F. Studer and B. Raveau, J. Phys. Chem. Sol. 44:389 (1983).

6. $LiTi_2O_4$: K.W.Ng. R.N. Shelton, E.L. Wolf, Phys. Lett. 110A:423 (1985).
   LiV_2O_4: B.L. Chamberland and T.A. Newston, Sol. St. Comm. 58:693 (1986).

7. J.B. Goodenough, Prog. Sol. St. Chem. 5: chapter 4 (1971); J. Sol. St. Chem. 12:148 (1975).

8. A. Auerbach and P.B. Allen, Phys. Rev. B 29:2884 (1984).
   D. Belitz and W. Schirmacher, J. Phys. C: Sol. St. 16:913 (1983).

9. M. Lehmann, C. Nölscher, H. Adrian, J. Bieger, L. Söldner and G. Sae-mann-Ischenko, p. 107 in Supercond. in 'd- and f-band Metals' (4th Conf. of Series), ed.: W. Buckel and W. Weber, Nuclear Centre, Karlsruhe.

10. K.A. Muller, M. Pomerantz, C.M. Knoedler and D. Abraham, Phys. Rev. Lett. 45:832 (1980).

11. S. Waki, N. Kasai, S. Ogawa, Sol. St. Comm. 41:835 (1982).
    D. Carre, D. Avignant, R.C. Collins, A. Wold, Inorg. Chem. 18:1370 (1979).

12. G. Shirane, W. Tomlinson and D.E. Moncton, p. 381 in 'Superconductivity in d- and f-band metals' (3rd Conf. in Series), ed.: H. Suhl and M.B. Maple, Academic Press, N.Y. (1980).

13. J.J. Hauser, M. Robbins, F.J. DiSalvo, Phys. Rev. B 8:1038 (1973).

14. R. Collins, R. Kaner, P. Russo, A. Wold, D. Avignant, Inorg. Chem. 18:727 (1979).

15. L.F. Mattheiss, Phys. Rev. B 8:3719 (1973).

16. G.Y. Guo and W.Y. Liang, J. Phys. C: Sol. St. 19:995 (1986).

17. G. Thiele, M. Köhler-Degner, K. Wittman, G. Zoubeck, Angewand. Chem. Intl. 17:852 (1978).

18. V. Riede, H. Newmann, H. Sobotte, F. Lévy, Sol. St. Comm. 38:71 (1981).

19. H. Sugiura and T. Yamadaya, Sol. St. Comm. 49:499 (1984).

20. a) J.A. Wilson, J. Phys. F. Metals 15:591 (1985).
    b) J.A. Wilson, Phil. Trans. Roy. Soc. A314:159-177 (1985).
    c) R.L. Withers and J.A. Wilson, J. Phys. C: Sol. St. 19:4809-4845 (1986).

21. E. Caruthers, L. Kleinman and H.I. Zhang, Phys. Rev. B 7:3753,3760 (1973).

22. a) N.J. Doran and A.M. Woolley, J. Phys. C.: Sol. St. 14:4257 (1981); J. Phys. C.: Sol. St. 16:L675 (1983).
    b) J.A. Wilson, J. Phys. F Metals 15:591 (1985).

23. D.A. Browne and K. Levin, Phys. Rev. B 28:5049 (1983).

24. M. Kataoka, J. Phys. C.: Sol. St. 19:2939 (1986); Phys. Rev. B 28:2800 (1983).

25. N. Suzuki, A. Yamamoto and K. Motizuki, Sol. St. Comm. 49:1039 (1984).

26. E. Sato, K. Ohtake, R. Yamamoto, M. Deyama, T. Mori, K. Soda, S. Suga and K. Endo, Sol. St. Comm. 55:1049 (1985), see also ref. 20.

27. M. Ghedira, M. Anne, J. Chenavas, M. Marezio and F. Sayetat, J. Phys. C: Sol. St. 19:6489 (1986).

28. C. Maule, J.N. Tothill, P. Strange and J.A. Wilson, J. Phys. C: Sol. St. to be published (1987).

29. J.P. Pouget, H. Laurois, T.M. Rice, P. Dernier, A. Gossard, G. Ville-
    neuve and P. Hagenmuller, Phys. Rev. B 10:1801 (1974).
    G. Villeneuve, M. Drillon, J.C. Lauray, E. Marquestant and P. Hagen-
    muller, Sol. St. Comm. 17:657 (1975).
30. I.S. Jacobs, J.W. Bray, H.P. Hart, L.V. Interrante, J.S. Kasper, G.D.
    Watkins, D.E. Prober and J.C. Bonner, Phys. Rev. B 14:3036 (1976).
31. M. Pasternak, A.J. Freeman and D.E. Ellis, Phys. Rev. B 19:6555 (1979).
32. R. Podloucky, J. Phys. F: Met. 14:L145 (1984).
    K. Selte, L. Birkeland, A. Kjekshus, Acta Chem. Scand. A 32:731 (1978).
33. R. Vincent and R.L. Withers, to be published.
34. G.I. Makrovetskii and G.M. Shakhlevich, Phys. Stat. Sol. (a) 61:315
    (1980).
35. G. Münninghoff, W. Treutman, E. Hellner, G. Heger and D. Reiner, J.
    Sol. St. Chem. 34:289 (1980).
36. M. Lenglat, P. Foulatier, J. Dürr and J. Arsène, Phys. Stat. Sol. (a) 94:
    461 (1986).
37. H. Tanino and K. Kobayashi, J. Phys. Soc. Jap. 52:3978 (1983).
38. G. Thiele, M. Steiert, D. Wagner, H. Wochnes, J. Anorg. Allg. Chem.
    516:207 (1984).
39. P. Day, C. Vettier and G. Parisot, Inorg. Chem. 17:2319 (1978).
40. J.A. Wilson, Structure and Bonding 32:57-19 (1977).
41. J.A. Wilson, F.J. DiSalvo and S. Mahajan, Adv. Phys. 24:117 (1975).
42. R.L. Withers and J.A. Wilson, J. Phys. C: Sol. St. 19:4809 (1986).
43. H. Rietschel and L.J. Sham, Phys. Rev. B 28:5100 (1983).
    P.B. Allen and B. Mitrovič, Sol. St. Phys. 37:2 (1982).
44. R. Brusetti, J.M.D. Cory, G. Czjzek, J. Fink, F. Gompf and H. Schmidt,
    J. Phys. F Metals 10:33 (1980).
45. P. Hertel and J. Appel, Phys. Rev. B 33:2098 (1986).

SOME RECENT EXPERIENCES IN NARROW-BAND PHENOMENA:

URu_2Si_2 AND $YBa_2Cu_3O_7$

J.A. Mydosh

Kamerlingh Onnes Laboratorium
der Rijks-Universiteit Leiden
2300 RA Leiden, The Netherlands

In a series of recent experiments we, at Leiden, have been focussing our materials-research program upon two different types of systems, namely the heavy fermions represented by URu_2Si_2 and the oxidic high-T_c super-conductors, e.g. $YBa_2Cu_3O_7$ and various substitutions.

Considering first the former intermetallic compound, our experimentation [1], especially neutron scattering [2], has proven the coexistence of a magnetically-ordered and superconducting ground state. This observation completes the four possibilities for the ground state of a heavy-fermion material: (i) remaining a heavy Fermi liquid down to the lowest temperatures ($\approx 10mK$), e.g. the "vegetables" $CeAl_3$ and $CeCu_6$, (ii) undergoing a superconducting transition, e.g. $CeCu_2Si_2$, UBe_{13} and UPt_3, (iii) developing a magnetically ordered (antiferromagnetic with reduced moments) state, e.g. U_2Zn_{17} and UCd_{11}, and finally (iv) the previously-mentioned coexistence between (ii) and (iii) as is the case for URu_2Si_2.

Theory offers no guidance as to what new materials are heavy fermions and, if so, whether they will attain a certain of the above ground states. Consequently one must employ an emperical approach, i.e. look at many different systems and follow their trends. It would be of great help to materials researchers if there were certain "guiding principles" or "basic ingredients" which could be established to predict the favorableness of heavy-fermion behavior. What are the proper elemental combinations for this phase as viewed not only from the 4 or 5f elements but also from the transition metal and the "simple" components such as Be, Cu or Al and Si or Ge?

A second problem here lies with the heavy-fermion magnetism. In the past the unusual superconductivity has attracted the most attention. But with the new group-theory classification schemes for the type of super-conductivity and some theoretical insights into the mechanism, e.g. a Cooperpair formation via exchange of antiferromagnetic spin fluctuations, progress is nicely being made. Nevertheless, the best or even a good description for the magnetism has not yet been formulated. U is a complex element since it can take on various (3) valences in a metal and thereby have widely different effective moments. In addition the importance of crystal-field effects are not as well established as in the rare-earth compounds. Slowly the importance of antiferromagnetism, either as fluctuations, UPt_3, or ordering, U_2Zn_{17}, is being appreciated as a general property of the heavy-fermion state. However, the reduced moments pose the questions how and to where have they disappeared. And then at low temperatures how and through what means do the surviving moments form their antiferromagnetically ground state? Is a SDW favorable and all we need? These queries are difficult to answer from the theoretical standpoint. Afterall the Kondo lattice and coherence has been around for many years. Further, itinerant magnetism and SDW are from the distant past with Cr being the archetypal example. At present more interest should be placed on the magnetism per se and as a novel mechanism for the heavy-fermion superconductivity. Perhaps here URu_2Si_2 can play an enlightening role especially since there is such a drastic reduction in the magnetic moment and the (anisotropic) superconductivity occurs above 1 K.

This brings us to the final problem of our heavy-fermion ground state. How does the coexistence in URu_2Si_2 come about? The size of the discontinuity in the specific heat at the superconducting transition and the large value of the initial slope of the upper critical field indicate that the superconductivity is carried by the heavy (hybridized) 5f electrons. However, these same electrons were necessary to create the SDW with partially-gapped Fermi surface. Hence, a most unusual behavior results in which one branch of the Fermi surface causes the magnetic ordering while yet another is available for Cooper pairing and a superconducting phase. This symbiosis is most certainly aided by there being a large density of states at E_F and the dominance of antiferromagnetic interactions. Additional theory [3] and band-structure calculations are warranted to understand the details of this remarkable coexistence.

We now jump over to briefly consider the oxidic superconductors and pose the question: where is the superconductivity. Probably by the time these Proceedings appear, this question would be fully answered. Nevertheless, the present state of affairs is controversial and the situation remains unclear. Indeed, a totally new class of superconductors has been discovered with $La_{2-x}Sr_xCuO_4$ ($T_c \approx 35K$) and $YBa_2Cu_3O_7$ ($T_c \approx 90K$). However, are these materials bulk superconductors? Here the difficulty is with the glassy behavior [4] of the powder samples and their ceramic "bulk". Are the individual grains uniform (2-D?) superconductors, and are the interfaces formed by the sintering process the origin of the long-time metastabilities and the strong irreversibilities? Or, on the other hand, is there an intrinsic cause within the grain, e.g. defects, dislocations, stacking faults, the 2-D nature, anisotropies etc.? One path towards a solution lies with avoiding the powder preparation methods and fabricating new types of samples as single crystals where the defect structure is known or thin films, epitaxially grown. In the meantime and if sufficient patience is at hand the glassy dynamics offers a unique probe for investigating these superconductors and for learning more about their true nature.

Another extraordinary feature of these superconductors is their unaffectedness to replacing the Y with a magnetic rare earth. In one of our recent experiments we fabricated a high-quality sample of $GdBa_2Cu_3O_7$ (various dilutions of Gd are presently being studied). Its normal state and superconducting transition were similar to the best Y-specimens [5]: $\rho_n(95K) \approx 300\mu\Omega$-cm, $d\rho/dT = 2.2\mu\Omega$-cm/K, $T_c(50\%) = 93K$, $\Delta T_c(10\%-90\%) = 1.5K$ and a large (3×10^{-2}emu/g below 40K) diamagnetic susceptibility beginning at 92.5K. The only difference for the Gd sample was that a very small but distinct peak appeared on the negative susceptibility plateau at 2.3K. Such an effect suggests the occurrence of an antiferromagnetic transition within and disconnected from the superconducting state, nota bene $\rho=0$ down to at least 1.2K. So once again it would seem that a remarkable coexistence of magnetic order and superconductivity takes places, but now in different atomic layers of the compound. This decoupling of the two phases opens the whole rare-earth (and actinide?) series for fabricating novel and unusual oxidic superconductors. We close by asking how many new high-T_c superconductors will be found in this oxide (or ??) class of materials.

References

1. T.T.M. Palstra, A.A. Menovsky, J. van den Berg, A.J. Dirkmaat, P.H. Kes G.J. Nieuwenhuys, and J.A. Mydosh, Phys. Rev. Lett. 55, 2727 (1985).
2. C. Broholm, J.K. Kjems, W.J.L. Buyers, P. Matthews, T.T.M. Palstra, A.A. Menovsky, and J.A. Mydosh, Phys. Rev. Lett. 58, 1467 (1987).
3. See M. Kato and K. Machida, to be published.
4. K.A. Müller, M. Takashige, and J.G. Bednorz, Phys. Rev. Lett. 58, 1143 (1987).
5. R.J. Cava, B. Batlogg, R.B. van Dover, D.W. Murphy, S. Sunshine, S. Zahurak, and G.P. Espinosa, Phys. Rev. Lett. 58, 1676 (1987).

VALENCE FLUCTUATIONS IN $La_{2-x}Sr_xCuO_4$

R.A. de Groot

ESM, Faculty of Sciences

Toernooiveld, 6525 ED Nijmegen, the Netherlands

INTRODUCTION

One of the interesting phenomena occurring in narrow band materials are valence fluctuations. We address this problem for the case of high T_c superconductors[1,2,3] here. The characteristic feature of the recently discovered high-T_c superconductors derived from the oxide La_2CuO_4 is the mixed valence of the copper constituent (Cu^{2+} and Cu^{3+}), which results from substituting a fraction of the trivalent La-ions by bivalent Ca, Ba or Sr dopants. A similar mixed valence is also found in the two previously discovered oxide superconductors $BaPb_{1-x}Bi_xO_3$ (Bi^{3+}, Bi^{5+})[4] and $Li_xTi_{2-x}O_4$ (Ti^{3+}, Ti^{4+})[5].

An important question is whether the mixed valence is static, namely certain Cu-sites are bivalent and others trivalent, or whether all the Cu sites are equivalent and have a dynamically fluctuating charge. To compare, in the monoclinic phase of $BaPb_{1-x}Bi_xO_3$ there are crystallographically inequivalent Bi-sites, and these have been interpreted as having different valence. However, infrared and X-ray photoemission spectroscopy[6] give no indication of a significant charge difference at these sites.

The nature and the role of the mixed valence of Cu in the new compounds have not been established. In most papers which have been circulating recently and which suggest a theoretical explanation, the only effect of the transforming some of the Cu^{2+} to Cu^{3+} is to suppress the tetragonal or orthorhombic transition leading to the semiconducting ground-state of La_2CuO_4. This is achieved by destroying the nesting required for a Peierls transition[7], by shifting the Fermi energy away from a Van-Hove singularity to avoid a band Jahn-Teller effect[8], by "melting" a resonating valence-bond ground-state[9], or by adding a bipolaron condensate to a commensurate charge density wave[10].

In the present study we demonstrate that the doping of the La_2CuO_4 compound produces dynamical valence-fluctuations on the Cu-sites and we propose that these charge fluctuations play an important role in the mechanism of superconductivity.

SUPERCELL BAND CALCULATIONS

In dealing with mixed valence systems or valence fluctuations a similar problem arises as in magnetism: the choice of the unit-cell. If we calculate the electronic structure of chromium in its primitive (bcc) unit-cell there are only two solutions possible: the non-magnetic and the ferromagnetic solution. This is because we have forced all chromium atoms to be equivalent. The calculation in the conventional (cubic) bcc structure allows three solutions: the two mentioned previously and an antiferromagnetic solution with alternating spins in the (111) direction. This latter solution is found to be the most stable, but it will be clear that this is not the magnetic structure observed experimentally. Hence the result of a calculation can depend on the choice of the unit-cell (a situation which makes the phrase ab-initio less meaningful in the case of a solid as compared with a molecule). It is clear that in order to allow valence fluctuations, we need a unit-cell containing more than one formula-unit; the so-called super-cell.

Valence fluctuations have been found before the band calculations in the case of $PbTaS_2$ $SnTaS_2$[11] and α-Ce[12]. Let us briefly discuss the experimental situation in $SnTaS_2$[13]. Photoemission spectroscopy shows a clear splitting of the Sn-core lines, indicating a mixed valence situation. Mössbauer spectroscopy shows one type of Sn only, but also indicates an unusual strong anisotropic thermal motion of the Sn atoms.

A supercell calculation on $SnTaS_2$ with all atoms at their equilibrium positions shows no mixed valence, even if the calculation is started with inequivalent Sn potentials. But a mixed valence solution is obtained in the calculation if a non-negligible distortion of the Sn-lattice is introduced, as suggested by the Mössbauer data. So there is an important difference with the magnetic case. In the calculation of a magnetic system one needs a small perturbation initially to avoid the system to be trapped in a meta-satable state (a calculation on Fe started with both spin-directions equivalently will remain non-magnetic). In the case of charge fluctuations we need a non-negligible symmetry breaking phenomenon to be present all the time. In the case of $La_{2-x}Sr_xCuO_4$ this symmetry breaking operation is present in the form of the actual substitution of La by Sr. **But this implies that it will not be possible to observe valence fluctuations even in a supercell on this material if one uses the virtual crystal approximation.**

DETAILS

In order to allow for the substitution of La by Sr we doubled the unit-cell by employing the primitive tetragonal unit-cell. This leads to a unit-cell of composition $La_3SrCu_2O_8$. The direct consequence of the replacement of La by Sr is the reduction of symmetry leading to two inequivalent Cu atoms in the unit-cell. In this sense our calculations differ from calculations employing the virtual crystal approximation. The introduction of Sr reduces the number of valence electrons from 82 to 81. A rigid band model would lead to a lowering of the Fermi-energy and consequently an emptying of states of primarily Cu-d and O-p character and hence to an average smooth increase of the valence of Cu on all sites. In our calculations a different phenomenon happens. The introduction of Sr in fact introduces strong valence-fluctuations on the copper-sites. The valence-fluctuations are so strong here that they prevented us from obtaining converged results in spite of the application of various type of mixing and using actual mixing as low as .003.

It is well-known that Cu is very sensitive to Jahn-Teller deformations and that Jahn-Teller deformations are sensitive to the valence of Cu[1]. The

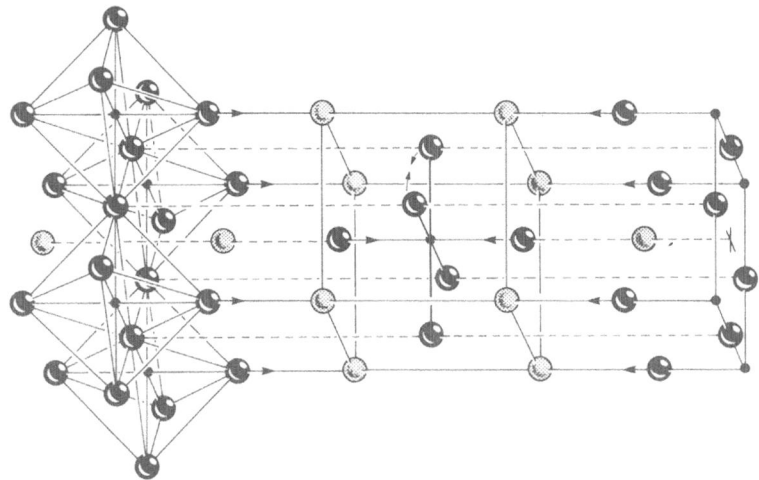

Fig. 1.   Unit-cell of La_2CuO_4. Bold arrows indicate additional Jahn-Teller
deformations applied to converge the calculation on $La_3SrCu_2O_8$.
Small arrows indicate the deformations taken into account in the
calculation for the peroxide-mode. The latter calculation did not
need the additional Jahn-Teller deformations.

undoped system shows already an elongation of the Cu-O distances in the z-
axis due to the JT-effect. Presuming that, conversely, the stability of the
static occurrence of a Cu atom in a certain valence would be sensitive to
JT-deformations, we attempted to help to converge the calculation by allowing
additional JT distortions.

For one type of Cu the JT deformations were removed by shifting the
Oxygens towards the Cu, for the other Cu the same displacement of O was
introduced into the opposite direction. The resulting deformations are
shown in Fig. 1. This calculation could be converged. The resulting Cu
partial density of states curves showed big differences as displayed in
Fig. 2. Also a difference as big as .33 electrons was found for the number
of d electrons within identical Cu-spheres. It is an important question
whether these differences, indicating different Cu valences, are due to the
introduction of Sr or to the additional JT deformations necessary to converge
the calculation. In order to investigate this question we performed a calculation on $La_4Cu_2O_8$ without Sr but with the same deformations of the oxygen-
lattice as in the previous calculation. The calculation converged smoothly
without any differences between the still inequivalent copper atoms whatsoever. We conclude that the partial substitution of La by Sr introduces
valence-fluctuations on the copper-sites.

THE DYNAMIC PEROXIDE COUPLING

The electronic structure of compounds consisting of metals and group
VI elements can be described usually in terms of a filled, broad anion band
and narrow transition-metal d states, with some degree of hybridization.
The position of the metal d-states is usually above the anion valence-band,
although some filled d-states frequently overlap the valence-band. The

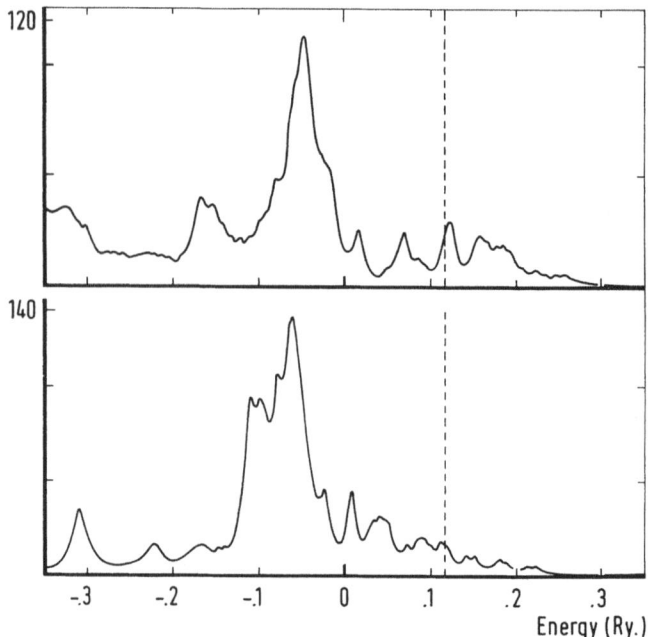

Fig. 2.   Partial density of states curves for the two inequivalent copper
atoms in the unit-cell in the case of $La_3SrCu_2O_8$ (smoothed with
a Gaussian, FWHM = .1 eV).

position of the d-states with respect to the valence-band is determined by
the electronegativity of the metal and its valence in the compound of interest. The deepest levels are found for noble metals in a high valence-
state. In such a case it can happen that empty d-levels start to overlap
the anion valence-band. This process leads to the creation of holes in this
anion-valence-band. A too large concentration of holes in the valence-band
is found to be always unstable, however. Energy can be gained by pairing
the anions. This leads to the splitting-off of an antibonding anion subband
above E_f and a subsequent reduction of the valences of both the anion and
the metal. Further consequences are the lifting of the anion-p transition-
metal-d degeneracy at the Fermi level and the restoration of an ionic-like
picture with the anions replaced by covalency bonded anion-pairs. It is
exactly this phenomenon which is responsible for the occurrence of the
pyrite, the marcasite and the peroxide-structures. The pyrite-structure,
for example, is simply the NaCl structure with the anion-positions occupied
by anion-pairs.

A similar situation occurs in the case $La_{2-x}Sr_xCuO_4$. The valence fluctuations introduced by Sr lead to a Cu^{3+} state, which is not stable with
respect to deformations of the anion-lattice. The difference from the usual
peroxide case is that the situation here is not static but since the valences
fluctuate dynamically, they lead to a dynamic coupling between the copper-
d electrons and deformations of the anion-lattice towards pair-formations.
This situation is exactly found in a calculation on $La_3SrCu_2O_8$ with the
required deformation of the anion lattice. These deformations are shown in
Fig. 1. The occurrence of the dynamic peroxide formation has the following
consequences.

1. A net charge transfer from oxygen to copper takes place. As a consequence the differences between the two copper-atoms are strongly reduced. The difference in d-electrons within identical spheres is reduced from .33 to .07 electrons.
2. It induces a split-off peak of primarily oxygen-p character just above E_f. This peak should be observable in BIS or X-ray absorption spectroscopy.
3. It leads to an asymmetrical broadening of the oxygen 2s states towards higher binding-energy, which should be observable in photoemission spectroscopy. The effect is however greatly diminished compared with static peroxides or pyrites since the oxygens are in the peroxide-state only a fraction of the time.
4. The dynamic peroxide formation leads to a depletion of states around the Fermi-energy.
5. It leads to the pairing of hole states on the oxygen. As a matter of fact it replaces single-electron states of primarily Cu-d and O-p character as charge carriers by oxygen-p hole pairs.
6. Structural consequences. Recently[14] it was found by neutron diffraction that $La_{2-x}Sr_xCuO_4$ has an orthohombic distortion between T_C and 180°K, and that there is an anomalous oxygen motion below T_C, where the system becomes tetragonal again. Although the exact nature of the oxygen anomalies was not determined we tentatively interpret these results as the occurrence of the dynamic peroxide mode, becoming static above T_C.

ACKNOWLEDGEMENT

Extensive discussions with A.R. Williams in all stages of this work are gratefully acknowledged. Part of this research was supported by the Stichting voor Fundamenteel Onderzoek der Materie (FOM).

REFERENCES

1. J.G. Berdnorz and K.B. Mueller, Z. Phys. B64:189 (1986).
2. C.W. Chu, P.H. Hor, R.L. Meng, L. Gao, Z.J. Huang, and Y. Wang, Phys. Rev. Lett. 58:405 (1987).
3. R.J. Cave, R.B. van Dover, B. Batlogg, and E.A. Rietman, Phys. Rev. Lett. 58:408 (1987).
4. A.W. Sleight, J.L. Gibson, and P.E. Bierstedt, Solid State Commun. 17:27 (1975); B. Batlogg Physica 126B:275 (1984).
5. D.C. Johnson, H. Prakash, W.H. Zachariasen and R. Vishwanathan, Mat. Res. Bull. 8:777 (1973).
6. J.Th.W. de Hair and G. Blasse, Solid State Comm. 12:727 (1973); A.F. Orchard and G. Thornton, J. Chem. Soc. Dalton Trans. 1238 (1977); L.F. Mattheis and D.R. Hamann, Phys. Rev. B28:4227 (1983).
7. H.B. Schuttler, J.D. Jorgensen, D.G. Hinks, D.W. Capone, and D.J. Scalapino, preprint.
8. J. labbe and J. Bok, preprint.
9. P.W. Anderson, Science 235:1196 (1987).
10. P. Prelovsek, T.M. Rice and F.C. Zhang, preprint.
11. H. Dijkstra, C. Haas, and R.A. de Groot, to be published.
12. R.A. de Groot, unpublished.
13. R. Eppinga, G.A. Sawatzky, C. Haas, and C.F. van Bruggen, J. Phys. C9:3371 (1976); R. Eppinga, G.A. Wiegers, and C. Haas, Physica 105B:174 (1981) and references therein.
14. D.Mc.K. Paul, G. Balakrishnan, N.R. Bernhoeft, W.I.F. David, and W.T.A. Harrison, Phys. Rev. Lett. 58:1976 (1987).

COMPOUND INDEX

SUBJECT INDEX

Kondo temperature, 2, 18, 20, 46, 70
Kondo effect, 2, 6-7, 17, 49
Kondo Hamiltonian, 17, 28, 74-6
Kondo resonance, 24, 26, 70, 169ff
Kondo volume collapse model, 18, 155ff

Lanthanides, 1, 5-6, 8-9, 49, 169
Lanthanide XPS, 8, 169
Lanthanide BIS, 8, 163ff
Lattice relaxation, 189
Local density approximation, 28ff, 74-7, 81ff, 98ff, 127, 133
Local spin density approximation, 82, 100
London criterion, 195,
Luttinger theorem, 20, 23, 83, 95

McMillan limit, 197
Magnetic fluctuations, 31
Magnetic instability, 26
Meisner effect, 196, 198
Measurement, perturbation by, 189
Measurement time scale, 20, 50, 56, 67, 167, 189
Migdal discontinuity, 24
Mixed valence, 5,6
Mode coupling, 32-33
Monte Carlo, 95
Mott-Hubbard insulator, 2, 4, 15-16, 20, 85, 97, 123, 142
Mott insulator, see Mott-Hubbard insulator
Mott transition, 4,15-16, 28, 123, 125
Multiplet Splittings, 5, 28, 86, 91, 99, 190
Narrow band
  definition, 1
  occurrence, 1, 195
Néel temperatures, 85, 97, 120-121

Orbital sizes, 3

Penetration depth, 195
Photoelectron spectroscopy, 145
Photoelectron satellites, 104
Point contact spectroscopy, 61-65, 70
  and local heating, 61-65
Polarization, 183ff, 190
Polaron, small, 202

Quasiparticles, 24, 27, 31, 67, 83, 86, 148

Rare earths, see lanthanides
RKKY, 2,16

Schrieffer-Wolff transformation, 75
SCF, 91
Screening mechanisms, 129, 149, 157, 166, 175ff, 183ff
Self-energy, 81, 84, 127, 141
Shake-up, 145, 175
Shell volume (4f), 3, 55
Singlet formation, 28
Slave Bosons, 170
Spin density functional, 24
Spin fluctuation temperature, 45-46
Spin fluctuation theories, 25, 45
Spin fluctuations, 19, 45-46
Spin linewidths, 55-58
Spin polarons, 114
Spin waves, 159
Stoner model, 2
Sudden approximation, 147, 189
Superconductivity, 16, 17, 46, 195ff
Superconducting oxides, 197ff, 201ff
Superconductors,
  type I, 195
  type II, 196,198
Superexchange model, 74-5, 97, 120

Tanabe-Sugano diagrams, 75
Time differential perturbed angular gamma ray distribution (TDPAD), 55
Transfer integrals, 2, 15-16, 118, 133ff, 205
Transition metals, 1, 4, 9, 55-56

U/W, 12, 100, 101

Valence fluctuations, 49
Valency, 5, 11

Wilson, 4

XAS, 51, 145, 179ff
XPS, 145ff, 155ff, 175